AF823313

Springer Texts in Business and Economics

Springer Texts in Business and Economics (STBE) delivers high-quality instructional content for undergraduates and graduates in all areas of Business/Management Science and Economics. The series is comprised of self-contained books with a broad and comprehensive coverage that are suitable for class as well as for individual self-study. All texts are authored by established experts in their fields and offer a solid methodological background, often accompanied by problems and exercises.

Byung T. Ro

The Geometry of Accounting

From Debits and Credits to Cartesian Coordinates

Springer

Byung T. Ro
Mitch Daniels School of Business
Purdue University
West Lafayette, IN, USA

ISSN 2192-4333 ISSN 2192-4341 (electronic)
Springer Texts in Business and Economics
ISBN 978-3-031-88891-5 ISBN 978-3-031-88892-2 (eBook)
https://doi.org/10.1007/978-3-031-88892-2

© The Editor(s) (if applicable) and The Author(s), under exclusive license to Springer Nature Switzerland AG 2025

This work is subject to copyright. All rights are solely and exclusively licensed by the Publisher, whether the whole or part of the material is concerned, specifically the rights of translation, reprinting, reuse of illustrations, recitation, broadcasting, reproduction on microfilms or in any other physical way, and transmission or information storage and retrieval, electronic adaptation, computer software, or by similar or dissimilar methodology now known or hereafter developed.
The use of general descriptive names, registered names, trademarks, service marks, etc. in this publication does not imply, even in the absence of a specific statement, that such names are exempt from the relevant protective laws and regulations and therefore free for general use.
The publisher, the authors and the editors are safe to assume that the advice and information in this book are believed to be true and accurate at the date of publication. Neither the publisher nor the authors or the editors give a warranty, expressed or implied, with respect to the material contained herein or for any errors or omissions that may have been made. The publisher remains neutral with regard to jurisdictional claims in published maps and institutional affiliations.

This Springer imprint is published by the registered company Springer Nature Switzerland AG
The registered company address is: Gewerbestrasse 11, 6330 Cham, Switzerland

If disposing of this product, please recycle the paper.

Preface

This book is about how geometry, and linear algebra, can be used to learn and study accounting. Geometry has long been used to explain concepts and theories in many fields of study. This, however, is not the case in accounting. Why is it so? The book is motivated by this very question. The book shows that the use of geometry along with linear algebra to learn and study accounting is possible. Traditionally, accounting has been learned and studied within the *accounting-only* frame, by relying exclusively on numbers and technical terms unique to accounting such as debit and credit, journal entry, T-account, journal, ledger, and so on. Little attention has been paid to the use of geometry and graphical tools, let alone linear algebra, to learn and study accounting.

The geometric approach adopted in this book is different from the traditional approach to accounting. It explores the features and principles of double entry accounting based on geometric concepts and theories. Although the use of geometry is common in many fields of study (e.g., mathematics, physics, engineering, economics, and finance, among others), such is hardly the case in accounting. The reason is not because it is not possible to use it in accounting but because no good effort has been made on the part of accountants, especially those in academia, to extend the domain of their thinking beyond the accounting-only frame.

Accounting transactions (events) have duality and size as their intrinsic attributes. Here, the duality means the dual sides, debit and credit, of accounting events, and the size means their monetary size, which is often called a "volume of transaction" or a "volume of trading." As such, accounting events have three intrinsic elements—debit, credit, and size, each representing a distinct dimension of the events. In this regard, they are multidimensional. From the perspective of geometry, each of those elements can be shown as a point or line on a distinct coordinate of a multidimensional geometric space. And an accounting event itself can also be shown as a point or line, or even as a space. Therefore, not only can accounting events be presented geometrically, but the features and principles of double entry accounting can also be explored geometrically.

From the perspective of geometry, the traditional approach to accounting is like a one-dimensional geometry. And the double entry system works as a two-way (debit-credit) classification method built on a single debit–credit dimension. When accounting events are recognized (recorded) in the one-dimensional setting, their multidimensional nature is suppressed or is not (or cannot be) made explicit. This is true especially for their size element. The size of accounting events is not the same as their debit and credit elements. It is double-sided and has a duality, whereas the debit and credit are single-sided and have no duality. Given this, the traditional one-dimensional approach has limitations when the nature and properties of accounting events as well as the features of double entry accounting are to be fully explored. A geometric approach, especially a multidimensional one, can mitigate such limitations and offer more insights.

The word *entry* in the name *double entry accounting* means a lot. It means that the essence of accounting is its entry, not exit, side. It is no wonder why there is no such thing as double *exit* accounting. The geometric approach in this book is concerned with the entry side of accounting. From a practical standpoint, the process of accounting consists of four sequential activities: measuring, recording, aggregating (summarizing), and reporting. Together, these activities constitute the main domain of accounting. They all are essential for the production and dissemination of accounting numbers, which is the main task of accounting. Of those activities, the recording step is at the heart of accounting. In a strict sense, the use of accounting numbers by users is not in the main domain of accounting as it belongs to the marketing or consumption of those numbers, so to speak. As for physical goods, producing the goods and marketing them is considered two distinct tasks so that the two activities are carried out by two different organizational branches, production department and marketing department. There is no reason why it should differ for accounting numbers even though these numbers are often characterized as the products of accounting.

Nonetheless, over the past half century or so, the production of accounting numbers, or more generally the entry side of accounting, has received far less attention than it deserves in the field of accounting. This is due to the so-called user-oriented approach in accounting, which is guided by the "decision-usefulness" view on the purpose of accounting and financial reporting. The geometric approach in this book is concerned with the production of accounting numbers, namely the entry rather than exit side of accounting.

The perspective of this book is markedly different from the perspective of the traditional approach to accounting. The readers, particularly graduate accounting students and accounting educators and scholars, who are interested in accounting beyond basics and have some knowledge of geometry and linear algebra, will likely find the book interesting. No book is complete, and this book is no exception. It is so, especially because the perspective of the book is so new in the field of accounting. If there is any error, it is the author's responsibility.

Torrance, CA, USA Byung T. Ro

Competing Interests The author has no competing interests to declare that are relevant to the content of this manuscript.

Contents

Introduction 1

1.1 Looking Beyond the Traditional Accounting Frame

The use of geometry to explain the concepts and theories is common in many fields of study such as mathematics, physics, engineering, economics, and finance. This, however, is not the case in accounting. Why is it so? This book is motivated by the very question. The book shows that geometry, and linear algebra, can be used to learn and study accounting. It does so by showing how to present or record accounting transactions (events) according to the double entry principle and explore the features and principles of double entry accounting in multidimensional geometric settings. In doing so, it uses geometric concepts such as point, line, space, and theories thereof and basic concepts and theories of vectors and matrices in linear algebra. As such, the approach in this book is different from the traditional approach to accounting.

Traditionally, the presentation of accounting transactions and the features and principles of double entry accounting have been learned and studied by relying exclusively on numbers and technical terms unique to accounting such as debit and credit, journal entry, T-account, journal, and ledger. In other words, accounting has been learned and studied within the accounting-only frame; no attempt has been made to look beyond the accounting-only frame and use geometric concepts or graphical tools to learn and study accounting, let alone vectors and matrices in linear algebra. Given this, the traditional approach is referred to as the *accounting-only* approach in this book.

From a practical standpoint, accounting—double entry accounting, to be exact—is a field of study on the method of recording accounting events and summarizing the entries of their debits and credits according to the double entry principle. This principle is a set of rules and procedures that guides how to recognize (record) accounting events based on their debit-credit duality and balance features or, more generally, the duality and balance features of the economic resources of an entity (a business entity or just a firm). The economic resources, referred to as capital resources hereafter, of a business entity or firm have a duality since they have dual

© The Author(s), under exclusive license to Springer Nature Switzerland AG 2025

B. T. Ro, *The Geometry of Accounting*, Springer Texts in Business and Economics,
https://doi.org/10.1007/978-3-031-88892-2_1

sides—source and use—as their intrinsic attributes. The source of the capital resources means where the resources come from, and the use means how they are held or used by the entity. Each side of the capital resources has a monetary value, and a monetary value common to both sides is the (monetary) size of the capital resources. It is this common monetary value that keeps the dual sides of the capital resources connected and hold them with balance.

As with capital resources, accounting events also have duality and balance as their inherent features. Here, accounting events mean not only arm's-length transactions but also any event that affects the capital resources of a business entity (firm). In accounting, the capital resources are shown in the name of assets, liabilities, and owner equity in a mathematical equation called the fundamental accounting model: assets = liabilities + owner equity. The left side of the model shows the uses of the economic (capital) resources by the firm as assets, and the right side shows their sources as liabilities and owner equity. Given this, accounting events include not only arm's-length transactions but also any event that affects the model in one way or another.[1] As with capital resources, they have a duality as they have dual sides. In double entry accounting, one of the dual sides is called *debit* (left) and the other *credit* (right). While the dual sides of accounting events typically mean their debit and credit, these two terms have a range of meanings in accounting, depending on how they are defined or used.[2] Regardless, they are always used as directional concepts in accounting.

Having a duality, accounting events always involve two things, one given (or received) on the debit side and the other received (or given) *in return* on the credit side. Or vice versa. Thus, they are recorded (recognized) by making a pair of entries—double entries, one for the thing on their debit side and the other for the thing on their credit side. The entry on the debit (credit) side is called a debit (credit) entry, and the account(s) shown in the debit (credit) entry is called a debit (credit) account(s). An account is a special device or place where the records of the same thing, whether given or received in accounting transactions, are entered and kept.[3]

For every accounting event, the thing on each side can be tangible or intangible. It has a monetary value, and a monetary value common to the things on both sides is

---

[1] In this book, accounting events (transactions) are defined in a broad sense. They include not only arm's-length transactions but also such events as natural events (e.g., earthquake, flood, or hurricane), the accrual of expenses and revenues, the depreciation of certain assets, and other events of similar nature. These events affect the assets, liabilities, and owner equity of a firm.

[2] For example, a debit means something given or received, a debit account, a debit entry, or a debit (monetary) amount on the left side of accounting events. It also means the left side of an account, a transaction, or a journal entry. And a credit means the opposite of each of those things above. Regardless, in accounting, the two terms are always used as directional concepts. A debit means the left side of accounting events, accounts, or journal entries, or the debit entry of accounting events. And a credit means the opposite of each of those. In this book, the two terms are used as directional concepts in a broad sense—debit meaning the left, and credit meaning the right, of any of those things above.

[3] If it is a cash account, for example, this account is a special device or place where only the records of cash receipts and cash payments are entered and kept.

realized as accounting events are executed between the two parties involved. This realized monetary value, called a "volume of trading" or a "volume of transaction," is the monetary size of accounting events. It is this monetary size that keeps the dual sides of accounting events connected and hold them with balance, just as the (monetary) size of capital resources of a firm does the same for their dual sides—source and use. Also, it is the same monetary size that keeps the debit entry and the credit entry balanced.

The recognition of accounting events based on their duality and balance features, along with the notion of debit and credit, is considered so unique. The two words *debit* and *credit* in particular are often understood as a pair of iconic terms for double entry accounting. And the duality and balance features of accounting events provide a conceptual basis for making double entries when accounting events are formally recognized (recorded). Or, more generally, they are the conceptual foundation of double entry accounting.

Given their duality and balance features, accounting events have three dimensions—debit, credit, and size (D, C, and S for short)—and three intrinsic elements—debit, credit, and size, denoted as d, c, and q in monetary terms. Each of these elements represents a distinct dimension of the events. In this regard, accounting events are multidimensional by nature. If any of those elements is absent, no accounting event can take place. For every accounting event, the three dimensions and the three elements d, c, and q always co-exist, co-vary, and do so with balance. Thus, if one of them is known or determined, the others are also known or determined. And a debit-credit balance is ensured for every accounting event and for every journal entry.

Under the traditional accounting-only approach, the multidimensional nature of accounting events is suppressed or not made explicit when the events are recognized under the double entry principle. This is true especially for their size element. While the debit and credit elements of accounting events are distinguished to some extent, the size element is either suppressed or assumed to be the same as the debit or credit element. This is unfortunate because the size of accounting events is not the same concept, nor the same in nature, as their debit and credit. The former represents the dual sides of accounting events and are hence double-sided, whereas the latter represent only one side (debit or credit) of the events and are hence single-sided. Also, while the debit and credit elements are a change (in monetary quantity)—either increase or decrease—in the debit or credit account(s), the size element is not such a change; it is just a "volume of trading" or a "volume of transaction." Yet, under the traditional accounting-only approach, the distinct nature of the size element of accounting events is not (or cannot be) made explicit when the events are recorded according to the double entry principle. This is particularly true for accounting events that have multiple accounts (entries) as a mix of increases and decreases on the debit or credit side (or both sides).

Besides, under the traditional approach, the three elements of accounting events are measured and shown only in monetary terms, although accounting events have a multitude of properties. Some of these properties are nonmonetary in nature and, hence, beyond monetary measurement and interpretation.

Given the above, it is difficult to gain good insights into the multidimensional nature of accounting events and the multitude of their properties under the traditional accounting-only approach. A geometric approach, especially a multidimensional one, can greatly mitigate much of the difficulties with the traditional approach.

1.2 The Use of Geometry in Accounting

As in many fields of study, it is possible to use geometric concepts, theories, and graphical tools to learn and study accounting. When viewed from the perspective of geometry, double entry accounting under the traditional accounting-only approach works as a one-dimensional (debit-credit) geometry (see Chap. 2 for details). A one-dimensional setting, however, does not conceptually fit for the multidimensional nature of accounting events. This is because it is not possible to show each of the three (debit, credit, and size) elements of accounting events on a distinct dimension. Yet, in a multidimensional setting, it is possible to do so. Each element of accounting events can be shown on a distinct dimension and done so according to the double entry principle. As such, it is possible to not only distinguish the debit and credit elements from each other but also make the size element explicit when accounting events are recorded under the double entry principle. Further, an accounting event itself can also be shown as a distinct point or line, or even as a space. As such, a multidimensional approach offers better (visual) insights into not only the multidimensional nature of accounting events but also the features of double entry accounting in general.

In a multidimensional geometric space, one dimension of the space is reserved for the debit entry (account) of accounting events, and the other for the credit entry (account). And still another dimension is reserved for the monetary size of the events. Therefore, a multidimensional setting fits conceptually well for the multidimensional nature of accounting events. While the size element of accounting events is equal in monetary terms to their debit or credit elements, it is not the same concept as their debit or credit element as indicated before. It represents the common monetary value of the two things interchanged in the accounting events, while neither the debit element nor the credit element does the same. As such, the former is double-sided and reflects the duality of accounting events, whereas the latter is single-sided and reflects no such duality. Besides, the size element of accounting events reflects a multitude of their properties, some of which are nonmonetary in nature and, hence, beyond monetary measurement and interpretation. Neither the debit element nor the credit element reflects such nonmonetary properties of accounting events.

Given the above, it is important to distinguish each of the three elements of accounting events from the others when the events are recognized according to the double entry principle. This is true especially for accounting events that have multiple accounts as a mix of increases and decreases on the debit or credit side, or on both sides. For such accounting events, the monetary quantity of the debit entries is not one continuous quantity but a multitude of discrete quantities. And some of these discrete quantities may be an increase, while the others may be a decrease. The same is also true for the monetary quantity of the credit entries. If so, the discrete monetary quantities of the multiple debit entries are not homogeneous because they are neither all increases nor all decreases. The same is also true for the monetary quantities of the multiple credit entries. Yet, under the traditional accounting-only approach, they all are shown on the same side, debit or credit, of the single debit-credit dimension. Adding up those increases and decreases on each side as total debits or total credits gives no meaningful aggregate; they are neither total increases nor total decreases.

As for the size of accounting events, it can be measured by adding the monetary amounts of the debit accounts or the credit accounts in absolute terms, whether these accounts are increases or decreases (or both). This can be done because, conceptually, the monetary size or volume of accounting events is an absolute scalar quantity. In fact, it is always a positive scalar quantity since no accounting event can take place with a zero or negative monetary amount. However, the monetary amount of the debit and credit entries, whether it concerns one account or multiple accounts, is not such an absolute scalar quantity; it is a directional *change*—increase or decrease in monetary quantity.

Unlike the one-dimensional geometric setting, in a multidimensional Cartesian geometric space with multiple coordinates, each element (d, c, and q) of accounting events can be shown as a point or line on a distinct coordinate according to the double entry principle. In a two-dimensional Cartesian space, for example, the debit element d of accounting events is a point or line on the horizontal (debit) coordinate, and the credit element c is a point or line on the vertical (credit) coordinate. The size element q cannot be shown on a distinct coordinate since there is no third coordinate for it in a two-dimensional space. Yet it can be shown as a point or line *off* the coordinates or even as a (flat) space. In a three-dimensional Cartesian space with a third coordinate, each element of accounting events can be shown as a point or line on a distinct coordinate. And an accounting event itself can also be shown as a distinct point or line *off* the coordinates, or as a (cubic) space.

Given the above, the use of a geometric approach, especially a multidimensional one, in presenting accounting events and exploring the features of double entry accounting has several merits. First, it fits conceptually well for the multidimensional nature of accounting events as indicated before. Thus, it helps gain good (visual) insights into the duality and balance features of accounting events and the multitude of their properties or, more generally, the features and principles of double entry accounting. This is not only because basic concepts and rich theories in geometry can be used to gain such insights but also because the elements of accounting events can be measured and shown in nonmonetary (geometric) as well as monetary terms. Being multidimensional by nature, accounting events have not only a monetary property but also a multitude of nonmonetary properties such as duality, force, power, quality, and informativeness, among others. These properties are beyond monetary measurement and interpretation. Under the traditional accounting-only approach, which relies exclusively on monetary measurement, it is difficult to gain insights into such nonmonetary properties of accounting events, let alone their multidimensional nature.

Second, the geometric presentation of accounting events may be considered a language of accounting just as accounting itself is often characterized as a language of business. When accounting events are recorded under the double entry principle, their debit, credit, and size elements are measured and shown as numbers. When shown in a geometric space, these numbers have a visual meaning. Each number represents a point, line, or space. And this point, line, or space has a meaning, which can be understood from the perspective of the double entry principle. What the number means is visually communicated by the point, line, or space. In this regard, the point, line, and space can be considered a means of communication—a language. It tells what the monetary size of the accounting event is, whether the debit or credit account increases or decreases, and how big this increase or decrease is. It also tells what the related journal entry to record the accounting event looks like. And so on. No verbal or written description is necessary to explain those things.

Third, a geometric approach focuses on the production, not use, of accounting numbers, which has not received much attention in the academic and research community in the field of accounting over the past half century or so. The geometric approach in this book focuses on the production of accounting numbers and the entry rather than exit side of accounting, which is at the heart of accounting. The term *double entry* or the name *double entry accounting* means that the essence of accounting is its entry, not exit, side. It is no wonder why there is no such thing as double *exit* accounting. From a practical standpoint, the process of accounting consists of the four sequential activities: measuring, recording (recognizing), aggregating (summarizing), and reporting. Together, these activities constitute the main domain of accounting. Of them, the recording step of the accounting process is at the heart of accounting. In a strict sense, the use of accounting numbers by users is not in the main domain of accounting. It belongs to the marketing or consumption of accounting numbers, so to speak. As for manufacturing firms, producing goods is the task of their production department, while selling them is the task of their marketing department. Such, however, has not been the case in accounting for many decades. The production of accounting numbers has received far less attention than it deserves, even if it is the main task of accounting and those numbers are often characterized as the products of accounting. This is due to the so-called decision usefulness view on the purpose of accounting and financial reporting, which was adopted by the accounting profession in the 1960s.[4] The geometric approach in this

[4]The decision usefulness view on the purpose of accounting was adopted by the American Accounting Association (AAA) in its position statement *A Statement of Basic Accounting Theory* (1966). In this statement, AAA takes the position that the purpose of accounting or financial reporting is to provide financial information useful to users and that the primary users of financial information are external rather than internal users. Ever since, the decision-usefulness view has become the conceptual foundation of the so-called user-oriented approach in accounting. It has set the direction of education, research, and textbooks in the field of accounting over the past half century or so.

book is concerned with the production rather than use of accounting numbers and the entry rather than exit side of accounting.

Finally, the use of a geometric approach can enhance the status of accounting as a practical field of study dealing with "bookkeeping" skills and "number-crunching" techniques to a level close to other fields of study where geometric concepts and theories are widely used. Mathematics, physics, economics, and finance are just some examples of such fields. In economics, for example, the determination of the price of a good as a function of its supply and demand is often explained graphically with points and lines in a two-dimensional geometric space. The maximization of personal utility subject to a budget constraint is also explained graphically in such a geometric space. In finance, the theory of portfolios, the theory of optimal investment decisions, and the mean–variance relation of investment returns respectively are often explained graphically. There is no reason why such cannot be done in accounting. Yet it is hardly the case in accounting. This is not because it is not possible to use geometric concepts and theories to learn and study accounting but because no good effort has been made on the part of accountants, especially those in academia, to extend the domain of their thinking beyond the traditional accounting-only frame.

Having such merits, a geometric approach can be a new avenue for learning and studying accounting. Not only can it help gain good (visual) insights into the features and properties of accounting events, but it also shows how double entry accounting in general can work geometrically. Further, it can enhance the status of accounting from a practical field of study to a level close to other fields of study where geometric concepts and theories are widely used.

1.3 The Content of the Book

This book consists of 10 chapters. Following Chap. 1, which briefly introduces the book, Chap. 2 examines the features of double entry accounting and the multidimensional nature of accounting events (transactions) in detail. It also discusses how double entry accounting works as a one-dimensional geometry under the traditional accounting-only approach. Chapter 3 extends what is discussed in Chap. 2 to a two-multidimensional (debit-credit) geometric setting. Using various types of accounting transaction examples, the chapter explains how to present accounting events in the two-dimensional setting according to the double entry principle. Chapter 4 extends what Chap. 3 discusses to a three-dimensional (debit-credit-size) setting using the same transaction examples. Chapter 5 examines two special (nonmonetary) properties—force and power, among others—of accounting events, which are not dealt with in the traditional accounting-only approach and in extant accounting literature. Chapters 6 and 7 respectively revisit what the previous three chapters discuss, based on concepts and theories in linear algebra, vectors and matrices in particular. In doing so, the chapters use the same transaction examples in the previous chapters. Chapter 8 discusses the aggregations of monetary measures in accounting in a two-dimensional geometric setting, given that aggregating those measures is part of the process of accounting. Chapter 9 examines financial ratios from the perspective of geometry. Although generating financial ratios is not a task of accounting, the chapter examines them geometrically as the financial numbers used to generate them come from accounting. Chapter 10 concludes the book with a brief summary, followed by a discussion of possible extensions. These extensions are concerned with some additional issues that are not dealt with in this book and left out for future work.

Reference

American Accounting Association (AAA). (1966). *A Statement of Basic Accounting Theory*. American Accounting Association.

The Double Entry Accounting as One-Dimensional Geometry

2

Double entry accounting has interesting features, which are rooted in three basic concepts—duality, balance (parity), and dual (double) entry. The first two of these concepts are so fundamental and pervasive that they are present in almost everything in existence, including economic (capital) resources and accounting transactions (events). The third concept is not as pervasive but unique to accounting. It is invented to guide how accounting events should be recorded based on their duality and balance features as the conceptual foundation. This chapter examines the features of double entry accounting in view of those three basic concepts and does so from the perspective of geometry. It does so by showing how double entry accounting works as a one-dimensional (debit-credit) geometry as it does in the traditional accounting-only approach. The chapter ends with a discussion of the limitations of the one-dimensional geometry, in view of the multidimensional nature of accounting events and the duality and balance features of double entry accounting.

2.1 Duality and Balance: The Conceptual Foundation of Double Entry Accounting

Duality and Balance as Fundamental and Pervasive Concepts

As a basic concept, duality is a fundamental and pervasive property that is present in almost everything in existence, whether tangible or intangible. Due to its pervasive nature, the concept has a wide range of meanings and interpretations, depending on how it is defined or used. Regardless, it means two contrasting aspects or sides of a thing or property. The two sides are contrasting yet symmetric and equivalent. For example, a person has a duality in that he or she has two contrasting yet symmetric body sides, left and right. A similar example is also found in geometry. A square, for example, has a duality in that it has dual sides that are contrasting yet symmetric and equal. A circle also has a duality, in many respects; if it is cut in half along its diameter, the resultant two half circles are contrasting yet symmetric and equal.

© The Author(s), under exclusive license to Springer Nature Switzerland AG 2025

B. T. Ro, *The Geometry of Accounting*, Springer Texts in Business and Economics,
https://doi.org/10.1007/978-3-031-88892-2_2

Regardless, the essence of the concept is the same; duality means the two contrasting yet symmetric sides (or aspects) of the same thing, tangible or intangible. And the dual sides are equivalent.

In reality, the notion of duality is often expressed with a pair of contrasting and dichotomous words such as front and back, top and bottom, good and bad, in and out, left and right, positive and negative, increase and decrease, action and reaction, day and night, joy and sorrow, life and death, and even heaven and hell. The list goes on. Since each pair of such words has contrasting yet symmetric meanings, one of them alone has little significance. To all living beings, for example, life does not mean much if there is no death, or vice versa. Similarly, heaven has little significance if there is no hell, or vice versa. Given the nature of duality as such a fundamental concept, it is not surprising that the concept is a major subject even for research and study in some fields of study such as economics, mathematics, and physics.

As with duality, balance (parity) is also a fundamental and pervasive concept. It is present in everything that has a duality, regardless of whether the thing is a tangible object or an intangible phenomenon or property. Not only does an object or a thing have a duality, but its dual sides also co-exist, co-vary, and do so with balance. In this regard, the two concepts—duality and balance—are inseparable. It is the balance feature that keeps the dual sides of the object or thing connected and holds them with balance. Without it, no object or thing can stay firm or stable. Nor can an object or a thing exist in harmony with other objects or things. Given this, it is not difficult to see how things can exist or move in harmony with one another and why nothing stands alone in the universe.

Geometric Illustrations

What duality and balance as a pair of fundamental concepts mean may be explained graphically. Figure 2.1a presents the two concepts in terms of a pair of arrows. Each pair of arrows represents what the two fundamental concepts jointly mean. The two arrows in each pair have the same length but point in opposite or perpendicular directions. Each arrow represents one of the dual sides of an object or a thing. In the first figure, for example, the two arrows have the same length and point in the opposite direction of each other from the dot as the point of reference (origin). Yet they co-exist, co-vary, and do so with balance. Hence, what they jointly mean is consistent with what duality and balance as a pair of fundamental concepts jointly mean. In the third figure, the two double-headed arrows have the same length, bisect

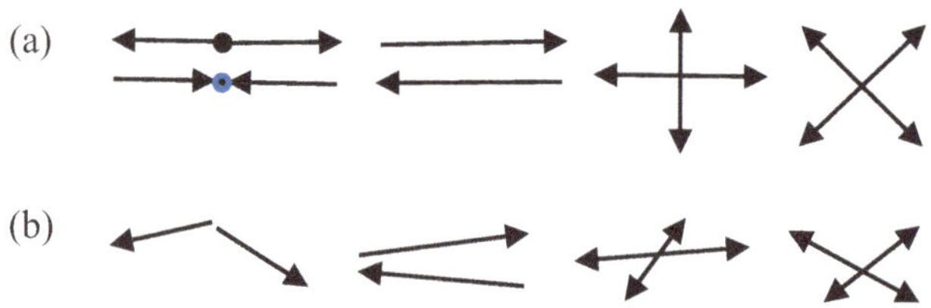

Fig. 2.1 The presentation of the concept of duality and balance with a pair of directed lines

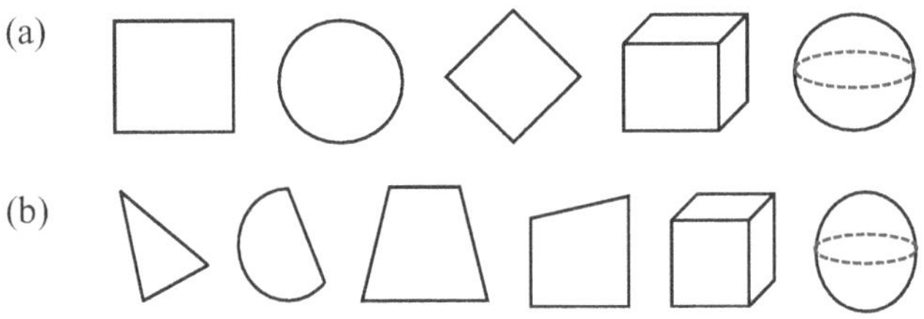

Fig. 2.2 The presentation of the concept of duality and balance as various geometric shapes

each other, and point in perpendicular directions. Together, they form an inter-locking, cross-like structure. This structure stands straight up, without tilting in any particular direction. Hence, what it shows is consistent with what duality and balance as a pair of fundamental concepts jointly mean. The other figures can be seen similarly.

In contrast, none of the figures in Fig. 2.1b shows the features of duality and balance and is hence consistent with what the pair of fundamental concepts jointly means. In the first figure, the two arrows not only differ in length but also point in different, but not opposite or perpendicular, directions. Hence, what they jointly mean is not consistent with what duality and balance as a pair of fundamental concepts jointly mean. In the third figure, not only do the two double-headed arrows differ in length, but they do not bisect each other. Nor do they point in opposite or perpendicular directions. Thus, what the inter-locking structure represents is not consistent with what duality and balance as a pair of fundamental concepts jointly mean. The other figures can be interpreted similarly.

As duality and balance can be illustrated with directed lines, so can they be with various geometric shapes. Figure 2.2a shows some geometric shapes that are consistent with what duality and balance as a pair of fundamental concepts jointly mean. The first geometric shape is a square. It has four identical faces, each having two pairs of contrasting yet equal sides. And the two adjacent sides of each face are equal and perpendicular. Next to it is a circle, which has many contrasting yet symmetric features. The diameter of the circle is the same no matter how it is measured. If the circle is cut in half along its diameter, the two resultant half circles are equal in size and symmetric in shape. This is true no matter how the diameter of the circle is measured and how the circle is cut in half. The third (rhombus-like) figure is also similar in this regard. Shown next is a cube, which has six faces that are the same in shape and equal in size. Each face is a square with two pairs of contrasting yet equal and symmetric sides. The round sphere also has similar features; its diameter is the same and the two resultant half spheres are equal in size and symmetric in shape, no matter how the sphere is cut in half and how its diameter is measured.

Unlike the various geometric shapes above, none of the shapes in Fig. 2.2b is consistent with what duality and balance as a pair of fundamental concepts jointly mean. The triangle, for example, is consistent with the notion of neither duality nor balance. It has no dual sides, the three sides are not equal, and the three interior angles are not the same. Nor is the next geometric shape consistent with what the pair of fundamental concepts jointly means. The next two quadrilaterals (trapezoids) have dual sides, but the dual sides are not equal. And, if they are cut through diagonally, the resultant two triangles are not the same in size, nor symmetric in shape, depending on how they are cut. Similar in this regard are the next two figures. Neither the rectangular cuboid (prism) nor the prolate spheroid is consistent with what duality and balance as a pair of fundamental concepts jointly mean. The cuboid has dual sides but these sides are not equal in length. If it is cut in half, the two resultant half cuboids differ in shape and size, depending on how it is cut. The same is also true for the spheroid; the diameter of the spheroid differs, depending on how it is measured. And, if the spheroid is cut along the diameter, the two resultant half spheroids differ in shape and size, depending on how it is cut in half.

2.2 The Features of Double Entry Accounting

Double Entry Accounting and Its Features

Double entry accounting is a field of study on the method of recording accounting events (transactions) and aggregating the monetary measures of individual accounts according to the double entry principle.[1] It is built on the duality and balance features of accounting events as the conceptual foundation. The method used to record accounting events and aggregate the monetary measures of individual accounts based on their duality and balance features is called the double entry system. And the set of rules and procedures used to do so is called the double entry principle. Conceptually, this principle is rooted in the duality and balance features of accounting events and capital resources (see Sect. 1.1, Chap. 1).

Under the double entry principle, every accounting event is formally recorded (recognized) by making a pair of entries—double entries, one on the debit (left) side of the accounting event and the other on the credit (right) side. The former is called the debit entry, and the latter the credit entry. What is duplicated in the dual entries is the monetary size (volume) of the accounting event. Together, the dual entries reflect the duality and balance features of the accounting event. Given the duality of accounting events, neither debit entry nor credit entry alone is complete and can reveal what the other side of the events is. Nor can it offer full insights into the duality and balance features of the events.

As a pair, the dual entries made to record an accounting event represent two things that are interchanged in the event between the two parties involved. One of the

[1] Contrasting to double entry accounting is what is called single entry accounting. In the latter, the duality (dual sides) of accounting transactions is not recognized when they are recorded. Instead, only one side of transactions is recorded by making a single entry rather than dual (debit and credit) entries. When goods are sold in cash, for example, the cash received is recorded as a single entry with no counterpart entry of the sales revenue realized. Besides, accounting events that are not arm's-length transactions, such as accruals and deferrals of revenues and expenses, are not recognized as they arise or until they are realized. The so-called cash basis of accounting is an example of single entry accounting.

things is given (or received) and the other is received (or given) *in return*. As such, each of them alone represents only one aspect or side of the accounting event and is unable to show what the other aspect or side of the event is. The thing on each side has a monetary value (quantity), and a monetary value common to the things on both sides is realized as the event is executed. This realized monetary value is the monetary size of the accounting event. It is this monetary size that is duplicated in the dual entries of the event and keeps the dual entries connected and balanced.

The double entry principle, or more generally double entry accounting, that guides how to record accounting events is known to have been invented in medieval Italy more than 500 years ago. While the specifics of the principle have been refined over time, the fundamentals of the principle have remained largely intact since then. Today, it is used worldwide and by nearly every kind of business entities, irrespective of the nature of the economic regime, capitalistic or socialistic, under which they operate. It can be understood uniformly across the borders of nations worldwide. In this regard, double entry accounting is a de facto global language of business.

Being such a pervasive language of business, double entry accounting has sailing features. These features are so interesting and unique that they make double entry accounting one of the finest record-keeping systems that the human intelligence has ever created. The most notable among the features of double entry accounting is that it recognizes the duality and balance features of both economic (capital) resources and accounting events. Given this, every accounting event is recorded by making dual (double) entries according to those features.

The capital resources of every entity (organization) including a business entity (firm) have a duality in the sense that they have dual sides—source and use. Here, the source means where the capital resources come from, and the use means how they are held or used by the entity. Given the duality, neither side alone of the capital resources can exist by itself. Nor can it reveal the other side. The dual sides are inseparable, always co-exist, and co-vary. Each side of the capital resources has a monetary value, and a monetary value common to both sides is the (monetary) size of the capital resources. It is this monetary size that keeps the dual sides of the capital resources connected and balanced. This balance feature is always preserved. There is no loss of value when the capital resources make a transition from their sources to their uses even if the process of transition takes time.

As with capital resources, accounting events also have duality and balance as their fundamental features. Accounting events have a duality in that they have dual sides—debit (left) and credit (right). As such, they always involve two things, one given (or received) by one party on one side and the other received (or given) *in return* by another party on the other side.[2] As with the dual sides of capital resources,

[2]One of the parties involved may be a person, an entity, or even Nature. If, for example, an accounting event occurs as a natural event such as earthquake, flood, or hurricane, Nature is considered one of the parties involved. See the definition of an accounting event in a footnote in Chap. 1.

the dual sides of accounting events always co-exist and co-vary. Hence, neither side alone can exist by itself. Nor can it reveal the other side.

Besides the duality feature, accounting events also have a balance feature as capital resources do on their part. Not only do the debit and credit sides of accounting events co-exist and co-vary, but they do so with balance. The dual sides of accounting events represent the two things interchanged, one given and the other received, between the two parties involved. The thing on each side has a monetary value, and a monetary value common to the things interchanged is realized as the events are executed. This realized common monetary value is the monetary size of the accounting events. It is this monetary size that keeps the dual sides of the accounting events connected and balanced, just as the monetary size of capital resources does for their dual sides.

In accounting, the notion of duality and balance relating to accounting events is implemented in the name of double entry, or just journal entry, when accounting events are recorded. This unique term literally means a pair of entries, one on the debit side of accounting events and the other on the credit side. What is duplicated in both entries is the monetary size of the events. Since this is true for every accounting event, the debit-credit balance feature is ensured not only for individual accounting events but also for all accounting events combined. The financial statements, on which all the accounts in use by a business entity are reported in summary, are also prepared according to the duality and balance features of accounting events and double entries. The same is also true for the capital resources of a business entity. The duality and balance features of capital resources are shown by a mathematical equation called the fundamental accounting model: total assets = total liabilities + total owner equity. This model shows not only the duality and balance features of capital resources but also the duality and balance features of all individual accounts and their monetary measures as ultimate aggregates.

Having such duality and balance features, accounting events are considered to have three dimensions—debit, credit, and size—(D, C, S) for short, represented by their three intrinsic elements—debit, credit, and size—(d, c, q) in monetary quantity. Each element represents a distinct dimension of accounting events. In this regard, accounting events are multidimensional by nature. If any of the three dimensions or elements is absent, no accounting event is complete or can take place. Once an accounting event takes place, the three elements of the event are recorded (recognized) based on the duality and balance features of the event, namely by making dual entries, with the monetary size of the event duplicated in both entries. In this regard, the size element of accounting events is considered double-sided, reflecting the duality of the events, while the debit and credit elements respectively are considered single-sided, reflecting no such duality.

The Double Entry System: A Four-Way Classification System

Double entry accounting, which is built on the duality and balance features of capital resources and accounting events, is implemented in the name of the double entry system. This system is designed to reflect the duality and balance features of accounting events as its conceptual foundation. In the double entry system, accounting events are recorded by distinguishing their dual sides (aspects), debit and credit, and making the entry on each side explicit as an increase or a decrease in monetary quantity. As such, the double entry system is designed to be a four-way (debit, credit, increase, and decrease) classification system. Thus, when accounting events are recorded, not only are their dual sides made explicit as the debit and credit entries, but the entry on each side is also made clear as an increase or a decrease in monetary quantity.

The double entry system as a four-way classification method differs from the double entry system viewed under the traditional accounting-only approach. Under this approach, it is viewed as a two-way (debit and credit) classification method built on a single (debit-credit) dimension, namely one horizontal debit-credit coordinate axis. This single coordinate is divided into two sections, debit (left) and credit (right), with the point of origin in the middle of the coordinate. There is no distinct coordinate for the debit and credit respectively of accounting events. Nor is there a separate section of the single coordinate for the debit and credit entries respectively as an increase or a decrease. Thus, whether a debit or credit entry (entries) of an accounting event means an increase or a decrease in an account is not (cannot be) made clear on the single debit-credit coordinate.

Two Basic Conventions in Double Entry Accounting

Before proceeding further, it seems appropriate to briefly discuss two basic conventions that are implicitly adopted in accounting. They are not part of the features of double entry accounting. Nor are they openly discussed in accounting textbooks. In this book, one of them is referred to as the value conservation convention and the other as the nonnegativity-of-money convention. Understanding what these two conventions mean is important to understand what is discussed throughout the book.

The Value Conservation Convention This convention, related to the duality and balance features of capital resources and accounting events, means that no monetary value is lost when the capital resources of a business entity (firm) make a transition from their sources to their uses. Thus, once a firm has a given amount of capital resources, the monetary value of the resources is the same, whether it is measured on their source side or on their use side. The same convention also applies to accounting events. When the thing on the debit side of an accounting event is interchanged for another thing on the credit side, they both are assumed to have the same monetary value. That is, the monetary value of the two things interchanged is preserved during the course of the accounting event being executed. If, for example, a firm obtains a bank loan (capital resource) of $1 million and purchases a building with the money, both the building and the loan are recorded and reported at $1 million. No value is assumed to have changed when the money makes a transition from the loan to the building. This is so even if the two transactions do not take place instantaneously or simultaneously and the real (present) value of the money keeps on changing over time. As such, both the building and the loan are recorded and reported at $1 million.

In a strict sense, the value conservation convention is not a realistic assumption since the time value of money continues to change over time. As for capital resources, it takes time for them to make a transition from their sources to their uses, during which the time value of money changes. This can make the real value of the capital resources measured on their source side differ from the real value measured on their use side. Similarly, accounting transactions can take time to go through. As for the purchase of the building with the loan money, for example, each transaction may well take time to go through, and the two transactions may not take place instantaneously or simultaneously. Therefore, the real value of the loan money can differ from the real value of the building when the purchase transaction of the building is completed and recorded by the firm. Yet, in accounting, a difference in monetary value between the building and the loan is not recognized when the purchase transaction is formally recorded. Both the building and the loan are recorded and reported at $1 million. This is so even if there is a time gap between the two transactions and the time value of the money continues to change over time.

The Nonnegativity-of-Money Convention This convention means that the value of money is nonnegative in substance. Thus, the monetary value (quantity) of the things interchanged between two parties in an accounting event is assumed to be nonnegative. This is true even if the accounting event decreases the debit and credit accounts. If an asset account is credited as a decrease, this does not mean that the money paid or the monetary quantity decreased is intrinsically negative; it simply means that the monetary amount of the asset account decreases. The same is also true for a monetary quantity that is debited to a liability account. If a liability account is debited as a decrease, this does not mean that the money paid to decrease the liability is intrinsically negative. When an account is debited or credited as an increase or a decrease, this is a matter of presentation, not substance. The same is also true when an increase or a decrease in monetary quantity is shown as a positive (+) or negative (−) number.

The nonnegativity-of-money convention also applies to the monetary size of accounting events. The size element of accounting events, unlike their debit and credit elements (accounts), is not a change (increase or decrease) in monetary quantity; it is just a monetary volume, namely a "volume of trading," which is an absolute scalar quantity. There is no such thing as a positive or negative monetary volume of an accounting transaction. In fact, the monetary size of accounting transactions is always positive since no accounting transaction can take place even with a zero monetary volume.

The nonnegativity-of-money convention is consistent with the nature of quantity in general as a basic concept. Conceptually, quantity is considered nonnegative in substance. It cannot be intrinsically positive or negative. It is not like temperature, which can be intrinsically positive or negative (e.g., hot or cold). Nor is it like color, which can be intrinsically different (e.g., black or white). Rather, it is like distance, size, or space, which is not positive or negative in substance. The quantity of money is considered one such quantity. It is an absolute scalar quantity and cannot be intrinsically positive or negative. There is no such thing as a negative asset or a negative liability, or a negative expense or a negative revenue. Nor is there such thing as negative one dollar or negative ten dollars.

Consistently with the nonnegativity-of-money convention, in accounting, no negative number or negative sign (−) is used to indicate a decrease in, or a negative balance of, individual accounts. Instead, all decreases and increases are shown based on their directions—debit (left) and credit (right), not by positive and negative numbers or with positive (+) and negative (−) signs. So is the positive or negative balance of individual accounts shown based on their directions—debit and credit, depending on the nature of accounts (e.g., asset or liability). Similarly, additions and subtractions of monetary measures in accounting are also carried out based on directions, debit and credit, without using the positive and negative signs. This is also true for total assets, total liabilities, and total owner equity as ultimate aggregates of all monetary measures in accounting when they are shown in the fundamental accounting model, which is discussed next.

2.3 The Fundamental Accounting Model and Double Entry Features

The capital (economic) resources of a business entity (firm), and their duality and balance features, are shown by a mathematical equation known as the fundamental accounting model in accounting:

$$\text{Assets} = \text{Liabilities} + \text{Owner Equity}$$

The model shows the capital resources of a business firm in the name of assets, liabilities, and owner equity. In accounting, the three elements are known as the basic elements of the fundamental accounting model. The model not only shows the stock balances of those elements but also articulates their relationship. Or, more generally, it shows the duality and balance features of capital resources in terms of those basic elements and their relationship. The assets represent the use side of the capital resources of the business firm, and the liabilities and owner equity represent their source side. The model shows that the monetary value of assets equals the monetary value of liabilities and owner equity. How the three basic elements of the model should be shown in the model is guided by the double entry principle. As the uses of capital resources, assets should be shown on the left side of the model and, as the sources of capital resources, liabilities and owner equity be shown on the right side. And the positive balances of capital resources and their increases should also be shown accordingly. To preserve the duality and balance features of capital resources, total assets as their uses should always match total liabilities plus total owner equity as their sources. The same is also true even for their changes, whether these changes are increases or decreases.

When viewed from the perspective of double entry accounting, the fundamental accounting model has interesting features. First, the direction of the model has special meanings; the left side means debit and the right side means credit. The former is reserved for the uses of capital resources, namely assets and their positive balances and increases. And the latter is reserved for the sources of capital resources, namely liabilities and owner equity and their positive balances and increases. This is consistent with the directional meanings of the debit and credit in double entry accounting. Given the directional meanings of the model, none of the following is a correct form of the model, although all is fine mathematically: liabilities + owner equity = assets, assets − owner equity = liabilities, or assets − liabilities = owner equity.

Second, reflecting the duality and balance features of capital resources, the model requires the sources of the capital resources and their changes (increases and decreases) to match the uses of the capital resources and their changes. The model also requires these changes to be shown according to what its directions mean—increases (decreases) in assets on the left (right) side and increases (decreases) in liabilities and owner equity on the right (left) side. As such, what the model shows is consistent with how the individual accounts are recorded and their changes are shown under the double entry principle. In other words, what the direction of the model means for capital resources is consistent with what the double entry principle means for accounting events. In this regard, the fundamental accounting model is closely related to the double entry principle, and the two are inseparable. The model tells how to present the capital resources of a firm and their changes at a macro level, while the double entry principle guides how to record individual accounts and their changes at a micro level.

Third, given its directional meanings, the model can be viewed as a mathematical presentation of the four-way classification scheme—debit, credit, increase, and decrease, just as the double entry system is a similar four-way classification scheme. The model shows not only how to present the uses (debits) and sources (credits) of the capital resources of a firm but also how to record their changes as increases and decrease. It shows that they should be presented in a manner consistent with the way individual accounts and their changes are recorded in the double entry system; asset accounts and their increases (decreases) on the debit (credit) side and liability and equity accounts and their increases (decreases) on the credit (debit) side.

The relationship between the fundamental accounting model and the double entry system, both as a four-way classification system, can be explained using the tableau in Fig. 2.3. The duality dimension of the tableau represents the dual sides, debit and credit, of both capital resources and accounting events. The direction dimension represents increases and decreases in both capital resources and individual accounts. These specifications divide the tableau into four cell blocks. The monetary quantity of each cell block is denoted by d, c, b, or a. Each of these notations means an increase or a decrease in monetary quantity.

With those specifications, the tableau shows how capital resources and their changes, and accounting events and their debits and credits as changes, are presented according to both the fundamental accounting model and the doble entry system. The sources of capital resources and their increases are shown on the credit dimension of the tableau, and the uses of capital resources and their increases on the debit dimension. And a change, whether an increase or a decrease, on the source (credit)

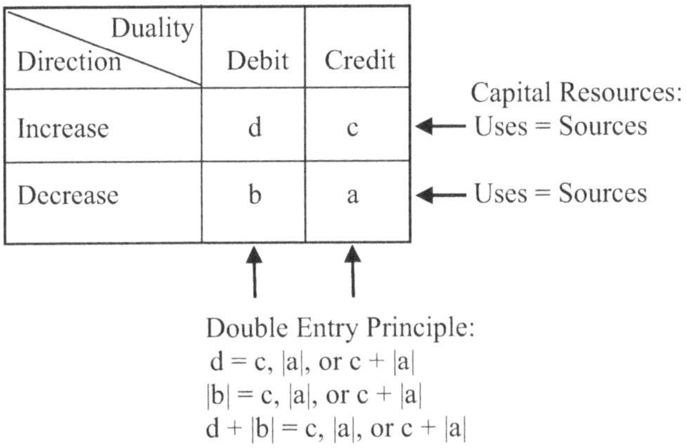

Fig. 2.3 The fundamental accounting model and the double entry system as a four-way classification method

side of capital resources should match a change on their use (debit) side. Consistently, when accounting events are recorded, an increase in the accounts relating to the use side of capital resources should be shown on the debit dimension, and an increase in the accounts relating to the source side of capital resources be shown on the credit dimension. And vice versa for a decrease in the accounts in each case.

For every accounting event, the debit entry as an increase or a decrease should match the credit entry as an increase or a decrease. The cell block denoted as d or b in the tableau means an increase or a decrease in the debit accounts, depending on the nature of these accounts (e.g., asset or liability). And the cell block denoted as c or a means an increase or a decrease in the credit accounts (e.g., liability or asset). As a decrease in monetary quantity, b and a can be shown as a negative number. This, however, does not mean that the decrease is intrinsically negative, as meant by the nonnegativity-of-money convention discussed before. Thus, it is shown in absolute value for notational convenience.

As shown in the figure, not only do the total sources of capital resources should match the total uses of the resources, but the debit of accounting events should also match their credit. This is true even for accounting events that have multiple accounts on the debit or credit side (or both sides). For every accounting event, therefore, the following hold according to the fundamental accounting model as well as the double entry principle:

$$d = c, |a|, \text{or } c + |a|$$

$$|b| = c, |a|, \text{or } c + |a|$$

$$d + |b| = c, |a|, \text{or } c + |a|$$

The first equation above means that an accounting event can take place with the debit d as an increase and the credit c as an increase, the credit |a| as a decrease, or both (c + |a|) in monetary quantity (absolute value). The second and third equations are interpreted similarly. Since both the fundamental accounting model and the double entry system have the same directional meanings, none of the above equations can be written in any other way. For example, the first equation cannot be written as c = d, |a| = d, or c + |a| = d. Nor can the second equation be written as c = |b|, |a| = |b|, or c + |a| = |b|. These are not acceptable in accounting although they are fine mathematically. Again, this is because the direction of the fundamental accounting model has special meanings, the left side as debit and the right side as credit.

2.4 Accounting Events and Their Types

An accounting event (transaction) is defined as an event that affects the basic elements (assets, liabilities, and equity) of the fundamental accounting model and their relationships. It can be not only an arm's-length transaction but also any event that affects those elements and their relationships in one way or another (see Sect. 1.1, Chap. 1). Given the definition, how accounting events affect the fundamental accounting model can be used as a basis for determining the types of accounting events. This can be explained based on the tableau of the double entry system in the previous section. Each pair of debit(s) and credit(s) in the tableau represents a particular type of accounting event. For example, an accounting event can be of the type [d c] or [d + b c] in journal entry form. When the types of accounting events are defined this way, there can be as many as four general types of accounting events shown in the journal entry form [debit credit]:

$$\text{Type 1: } \begin{bmatrix} d & c \end{bmatrix}, \begin{bmatrix} d & a \end{bmatrix}, \begin{bmatrix} b & c \end{bmatrix} \text{ or } \begin{bmatrix} b & a \end{bmatrix}$$

$$\text{Type 2: } \begin{bmatrix} d_1 & c \\ d_2 & \end{bmatrix}, \begin{bmatrix} d & c \\ b & \end{bmatrix}, \begin{bmatrix} d_1 & a \\ d_2 & \end{bmatrix}, \begin{bmatrix} d & a \\ b & \end{bmatrix}, \ldots$$

$$\text{Type 3: } \begin{bmatrix} d & c_1 \\ & c_2 \end{bmatrix}, \begin{bmatrix} d & c \\ & a \end{bmatrix}, \begin{bmatrix} b & c_1 \\ & c_2 \end{bmatrix}, \begin{bmatrix} b & c \\ & a \end{bmatrix}, \ldots$$

$$\text{Type 4: } \begin{bmatrix} d_1 & c_1 \\ d_2 & c_2 \end{bmatrix}, \begin{bmatrix} d & c \\ b & a \end{bmatrix}, \begin{bmatrix} b_1 & c_1 \\ b_2 & c_2 \end{bmatrix}, \ldots$$

where d, c, b, and a are all as defined before.[3]

Accounting events of type 1 have a single account (entry) on both debit and credit sides of the journal entry. The account in each entry can be an increase or a decrease. The sale of goods in cash is an example. This transaction is shown as [d c], with d for

[3]The four types of accounting events above are preserved in subsequent chapters of the book when accounting events are discussed or presented by their types.

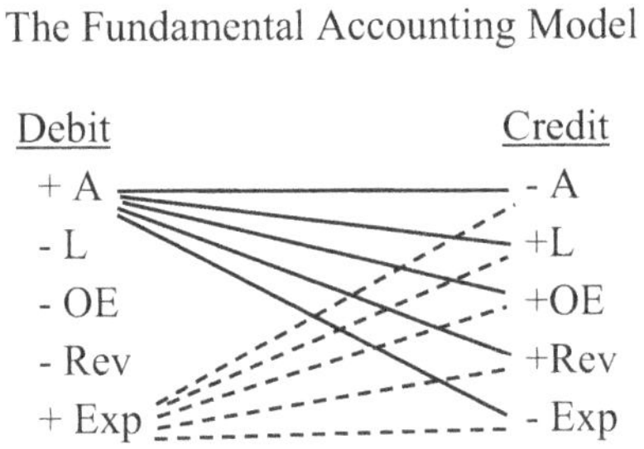

Fig. 2.4 Accounting events as combinations of the elements of the fundamental accounting model in journal entry form

the cash as an increase and c for the sales revenue as an increase. Accounting events of type 2 have multiple accounts (entries) on the debit side and a single account (entry) on the credit side. The multiple debit accounts may or may not all increase or decrease; one may increase and the other(s) may decrease. And the single credit account can increase or decrease. The sale of goods partly in cash and partly on account is an example. Accounting events of type 3 are the opposite of type 2 accounting events in that they have multiple accounts on the credit side. The purchase of goods partly in cash and partly on account is an example. More complex than those accounting events above are accounting events of type 4. Accounting events of this type have multiple accounts (entries) on both debit and credit sides. And the multiple accounts on each side may or may not change in the same direction; one may increase and the other decrease. The sale of a building with accumulated depreciation in cash at a gain is an example. This accounting event has cash and accumulated depreciation as the debit accounts and building and gain on sale as the credit accounts. As seen, type 1 accounting events are of the simplest type possible, and type 4 accounting events are of the most complex type. Since every accounting event has at least one debit account and one credit account, no accounting event can take place with debit accounts only or credit accounts only in double entry accounting. Thus, [d b] or [c a], for example, is not relevant since no such accounting event can exist or take place.

If an accounting event is of type 2, 3, or 4, the monetary quantity of the debit and credit entries respectively is not one continuous quantity but a multitude of discrete quantities: $d = d_1 + d_2 + \ldots + d_k$, $b = b_1 + b_2 + \ldots + b_k$, $c = c_1 + c_2 + \ldots + c_k$, and $a = a_1 + a_2 + \ldots + a_k$. The same is also true for the (monetary) size of the accounting event. For such accounting events, both the fundamental accounting model and the double entry principle require

$$\sum d_k = \sum c_k, \sum |a_k|, \text{or} \sum c_k + \sum |a_k|$$

$$\sum |b_k| = \sum c_k, \sum |a_k|, \text{or} \sum c_k + \sum |a_k|$$

$$\sum d_k + \sum |b_k|, = \sum c_k, \sum |a_k|, \text{or} \sum c_k + \sum |a_k|$$

Figure 2.4 shows all possible types of accounting events, in journal entry form, that can take place as a combination of the debit and credit elements of the fundamental accounting model. The element A stands for assets, L for liabilities, and OE for owner equity. The increase (decrease) in assets and expenses (Exp) is shown on the debit (credit) side in the figure. And the increase (decrease) in liabilities, owner equity, and revenues (Rev) is shown on the credit (debit) side in the figure. Thus, if a firm collects an account receivable in cash, for example, this transaction is shown by the combination of +A and −A.

As seen, each accounting event takes place as a combination of one or more debit elements of the fundamental accounting model and one or more credit elements. Each line, solid or dashed, connecting an element on the debit side and an element on the credit side represents a possible accounting event. For example, the line connecting +A and +L represents an accounting event of type 1 such as the purchase of an inventory asset on account, which increases the inventory asset as the debit entry and the account payable as the credit entry. Type 2, 3, and 4 accounting events can also be defined similarly, except that such accounting events are defined with a set of lines connecting the multiple elements on the debit or credit side, or both sides, of the model.

If the types of accounting events are defined as above, and based on the three basic elements A, L, and OE of the fundamental accounting model, there can be as many as 49 different types of accounting events that can take place.[4] This total includes not only arm's-length transactions but also other accounting events such as period-end adjustments, error corrections, and other similar types of accounting events that affect the basic elements of the model in one way or another. If the number of the elements of the model is expanded to five by separating Rev. and Exp from OE, the number of possible types of accounting events that can be written as various combinations of the debit and credit elements of the fundamental accounting model increases exponentially. And, if each of the five elements of the model is further decomposed into individual accounts (e.g., A into cash, account receivable, inventory, prepaid expenses, and so on), the total number of possible types of accounting events that can be defined as various combinations of the elements of the model rises beyond an easy calculation.

2.5 Double Entry Accounting in One-Dimensional Setting

The discussions thus far of the features of accounting events and the double entry system in this chapter suggest that accounting events have multiple dimensions (debit, credit, and size) and are hence multidimensional by nature. They also suggest that the double entry system, which is closely related to the fundamental accounting model, is a four-way classification system designed to make the debit and credit

[4]This number is obtained as follows: 1 debit and 1 credit (1:1) $\rightarrow 3 \times 3 = 9$, 1:2 $\rightarrow 3 \times 3 = 9$, 1:3 $\rightarrow 1 \times 3 = 3$, 2:1 $\rightarrow 3 \times 3 = 9$, 2:2 $\rightarrow 3 \times 3 = 9$, 2:3 $\rightarrow 3 \times 1 = 3$, 3:1 $\rightarrow 1 \times 3 = 3$, 3:2 $\rightarrow 1 \times 3 = 3$, and 3:3 $\rightarrow 1 \times 1 = 1$. Thus, $9 + 9 + 3 + 9 + 9 + 3 + 3 + 3 + 1 = 49$.

entries of accounting events explicit and show the individual accounts in those entries as increases and decreases.

Under the traditional accounting-only approach, such nature of accounting events and such feature of the double entry system are not made explicit when accounting events are recorded according to the double entry principle. Rather, the multidimensional nature of accounting events is suppressed, and the double entry system is treated like a two-way (debit and credit) classification method built on a single debit-credit dimension. In other words, double entry accounting works as a one-dimensional geometry under the traditional accounting-only approach. The single debit-credit setting, however, does not fit conceptually well for the multidimensional nature of accounting events. Nor does it fit for the double entry system as a four-way classification system.

In the one-dimensional setting, it is not possible to show each element (debit, credit, and size) of accounting events on a distinct dimension, while making it clear whether the debit and credit accounts (or entries) are an increase or a decrease (unless the name of the accounts is attached). On the single debit-credit dimension, a point is set as the point of origin (zero monetary quantity), and the left side of it is defined as debit and the right side as credit. An accounting event is then shown as a pair of points or lines on the single dimension, one on the debit side and the other on the credit side. As such, whether the point or line on each side is an increase or a decrease is not (cannot be) made clear.

Figure 2.5 gives an example of how the double entry system as a two-way classification method works in a one-dimensional (debit-credit) setting. In Fig. 2.5a, the point O is the point of origin (zero monetary quantity). The left side of it is reserved for the debit entry of accounting events, and the right side for the credit entry. Each side has the positive (+) *and* negative (−) signs, meaning that the entry (account) means an increase or a decrease on the same side. That is, both an increase in asset (or expense) accounts and a decrease in liability or owner equity (or revenue) accounts are shown as a point or line on the debit side. And both a decrease in asset (or expense) accounts and an increase in liability or owner equity (or revenue) accounts are shown as a point or line on the credit side. If, for example, an accounting event is of type 1 and increases the single account on each side, it is shown by the points d and c, which have the same distance from the point of origin. Or it is shown by the lines Od and Oc, which have the same length. And, if another type 1 accounting event of the same monetary size takes place but decreases the account on each side, it is shown by the points b and a—the same as d and c. Hence, d = b and c = a. And both d and b are shown on the debit side, and both c and a on the credit side, even if they are different kinds of changes. The two accounting events

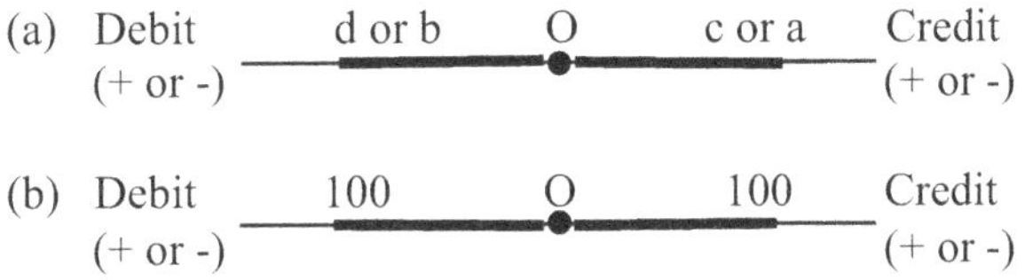

Fig. 2.5 The double entry system as a two-way classification method in a one-dimensional (debit-credit) setting

are shown by the same pair of points or lines on the same single dimension even if they are not the same. For both accounting events, their monetary size is suppressed or assumed to be the same as the point or line on the debit or credit side. Or it is ignored and, thus, not shown.

Figure 2.5b explains what is discussed above, with two simple transaction examples—the collection of an account receivable of $100 in cash and the payment of an account payable of $100 in cash. Both transactions are of type 1 and have the same monetary size. And the debit and credit accounts of both transactions have the same monetary measure. Thus, the two transactions are shown by the same pair of points or lines on the single debit-credit dimension. The monetary size of each transaction is suppressed as it cannot be shown on the single dimension. Both the cash $100 as an increase in the first transaction and the account payable $100 as a decrease in the second transaction are shown by the same point or line on the same debit side of the single line. The same is also true for $100 on the credit side of the single line. This is confusing or makes little sense. The confusion arises because the one-dimensional geometric setting does not conceptually fit for the multidimensional nature of accounting events and for the double entry system as a four-way classification system.

The accounting events in Fig. 2.5 are assumed to be of type 1, the simplest type possible, having a single account (entry) on both debit and credit sides. If an accounting event is of type 2, 3, or 4 and has multiple debit and credit accounts with a mix of both increases and decreases, presenting the event on the single debit-credit line is more difficult and can be more confusing. While the size element of the accounting events can still be suppressed, the multiple debit entries (accounts) cannot be shown by a set of points or lines, while still making it clear whether these points or lines mean an increase or a decrease in those accounts. The same is also true for the multiple credit entries (accounts). Both an increase in one account and a decrease in another are shown by the same point or line on the same side of the single debit-credit line if the two are the same in monetary amount. Again, this is confusing.

2.6 Limitations of the One-Dimensional Approach

Discussions in the previous section suggest that the single (debit-credit) dimension geometry does not conceptually fit well for the double entry system as a four-way classification method. Nor does it fit for the multidimensional nature of accounting events. More generally, it is not suitable for fully exploring the duality and balance feature of accounting events. This is because the geometric setting is one-dimensional. While the debit and credit elements d and c of accounting events are distinguished to some extent based on the point of origin on the single dimension, the size element q is suppressed. Or it is assumed to be the same as d or c even though it is not the same as the debit and credit elements. The size itself of accounting events is a distinct dimension and should be shown accordingly. It cannot be shown on the debit or credit dimension. It is double-sided as it reflects the duality

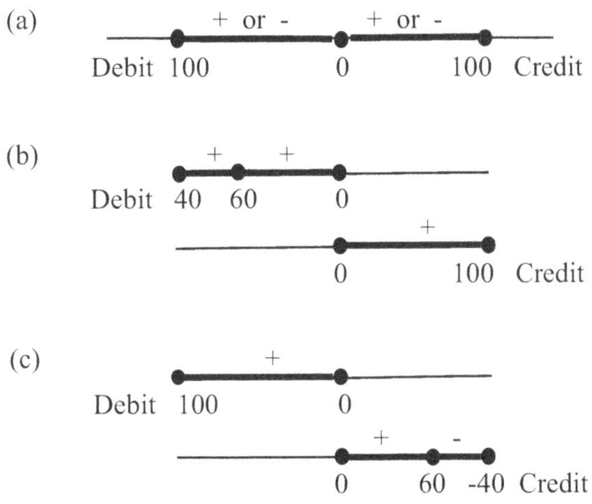

Fig. 2.6 Presenting two accounting events (sale and purchase) in a one-dimensional (debit-credit) setting

and balance features of the events, whereas the debit and credit of accounting events are single-sided and do not reflect those features of the events. Also, while the debit and credit of accounting events are a change, increase or decrease in monetary quantity, the size element is not such a change. When presented geometrically, therefore, q cannot (is not supposed to) be the same point or line as d or c on the same dimension.

What the above means is that the one-dimensional (debit-credit) setting has limitations when accounting events are presented geometrically. This is particularly true for type 2, 3, or 4 accounting events, which have multiple accounts as a mix of increases and decreases on the debit or credit side or on both sides. Given the complexity of such accounting events, it is difficult to present them on the single debt-credit geometric setting, while still making things clear.

The difficulties with the one-dimensional setting may be illustrated with two simple transaction examples of equal monetary size. Consider a firm that sells some goods for $100 in cash as one transaction and purchases other goods for $100 as another transaction. Given below are the journal entries to record the two transactions.

Cash	$60	
Account receivable	40	
Sale		$100
Purchase	$100	
Account payable		$60
Cash		40

As both transactions have the same monetary size, they have the same debit, credit, and size elements: $d = c = q = \$100$. The first transaction is a type 2 accounting event, and the second one is a type 3.

Figure 2.6 presents the two transactions on the single debit-credit dimension (line), considering double entry accounting as a one-dimensional geometry and the double entry system as a two-way classification method. In Fig. 2.6a, both transactions are shown by the same pair of points or lines. And the size element q of the transactions is not (cannot be) shown since there is no separate dimension (or section) for it. As seen, it is not clear how the two accounting transactions differ as they both are shown by the same pair of points or lines on the same (debit-credit) dimension.

The above difficulty is somewhat mitigated if the single debit-credit line is reset as two separate lines. In Fig. 2.6b, which presents the sale-of-goods transaction above, the debit entry has two components (60 and 40), both as an increase. Thus, they are shown as a pair of points or lines on the debit side of the single line. In Fig. 2.6c, which presents the purchase-of-goods transaction, the credit entry has two accounts. But one of them is an increase, shown by the point 60, and the other is a decrease, shown by the point −40. Although different in nature, both of these changes are shown on the same credit side of the single debit-credit line. This makes little sense.

The graphical illustrations above suggest that, while presenting the two transactions on the two different lines helps see their differences to some extent, it is still not enough to gain good insights into the nature of the two transactions and their differences. This difficulty can be greatly mitigated in a multidimensional geometric setting. In this setting, each element of accounting events is shown on a distinct dimension. And each dimension has positive and negative sides respectively which makes it possible to show each of the debit and credit accounts as an increase or a decrease. As such, the difficulty with the one-dimensional setting is greatly mitigated. This is true particularly for accounting events of types 2, 3, or 4, which have multiple entries (accounts) on the debit or credit side (or both sides).

Besides the difficulties seen above, the traditional one-dimensional geometry of double entry accounting has another difficulty; things including the size (q) of accounting events are all measured and shown only in monetary terms. The monetary measurement alone may work well if accounting events have a single property that can be measured and shown in monetary terms. Unlike the debit and credit entries, however, accounting events have a multitude of properties, some of which are nonmonetary and, hence, beyond monetary measurement and interpretation. Such properties include duality, force, power, quality, and informativeness, to name a few. Under the traditional one-dimensional approach, and with things measured and shown only in monetary terms, it is difficult to gain insights into such nonmonetary properties, as well as the multidimensional nature, of accounting events.

The Double Entry Accounting in Two-Dimensional Setting

3

The discussion of double entry accounting in the one-dimensional (debit-credit) setting in Chap. 2 can be extended to the case of a two-dimensional setting. This geometric setting is a two-dimensional Cartesian space. It is more consistent with the multidimensional nature of accounting events and the duality and balance feature of capital (economic) resources. It is also more consistent with the features of the double entry system as a four-way classification method. Accounting events are multidimensional as they have three distinct dimensions—debit, credit, and size, denoted as D, C, and S for short. In the two-dimensional (debit-credit) setting, the dimension for S is suppressed since there is no third dimension for it. Yet it can still be shown explicitly. This chapter discusses how to present accounting events and explore the features of double entry accounting in the two-dimensional (debit-credit) geometric space. It does so with four different types of transaction examples.

3.1 The Two-Dimensional Geometry of Double Entry Accounting

The Basic Setting

The two-dimensional geometric setting for double entry accounting is a two-dimensional Cartesian geometric space with two coordinates, one for the debit of accounting events and the other for the credit. This space has a point of origin (zero monetary quantity)—the point at which the two coordinates intersect each other. The two coordinates are mutually perpendicular (orthogonal). Since there is no third dimension (coordinate), the size dimension of accounting events is suppressed. In the two-dimensional (debit-credit) Cartesian space, the debit and credit elements d and c of accounting events can be shown as a point or line on the respective debit and credit coordinates. Although the size element of accounting events cannot be shown on a distinct coordinate, it can be shown as a distinct point or line, or even as a space (flat area).

© The Author(s), under exclusive license to Springer Nature Switzerland AG 2025

B. T. Ro, *The Geometry of Accounting*, Springer Texts in Business and Economics,
https://doi.org/10.1007/978-3-031-88892-2_3

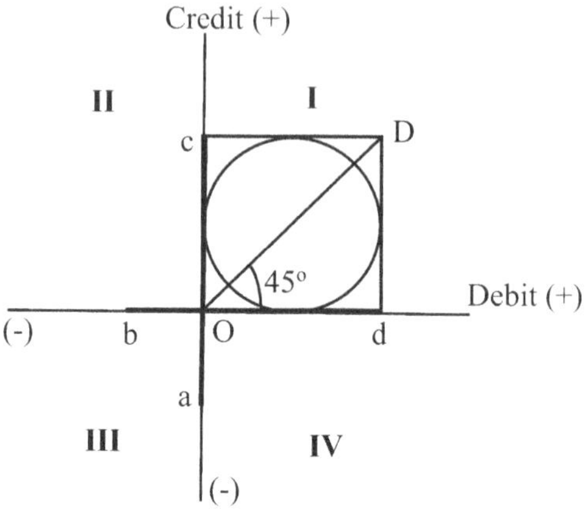

Fig. 3.1 A two-dimensional (debit-credit) setting for the presentation of simple-type accounting events

With the point of origin in place in the two-dimensional space, the point or line representing the debit and credit elements of accounting events is an ordered array, not just an array, of real numbers. So is the point, line, and space representing the size element q of accounting events. The point or line for the debit element d lies on the debit coordinate of the two-dimensional space, and the point or line for the credit element c lies on the credit coordinate. The size element q is also a point or line. Unlike the point or line for d and c, however, this point or line lies *off* the debit and credit coordinates.

Figure 3.1 gives an example of a two-dimensional Cartesian geometric setting where accounting events can be presented according to the double entry principle. The "Debit" and "Credit" coordinates represent the debit and credit dimensions respectively of accounting events.[1] Each coordinate has two sides, positive (+) and negative (−), and the point O at which the two coordinates intersect is the point of origin—the point of zero monetary quantity. With this point set, the two coordinates divide the entire Cartesian space into four quadrants, QI to QIV. The positive side of the debit coordinate is reserved for the debit entries of asset and expense accounts as increases, and the negative side for the debit entries of liability, owner equity, and revenue accounts as decreases. Consistently, the positive side of the credit coordinate is reserved for the credit entries of liability, owner equity, and revenue accounts as increases, and the negative side for the credit entries of asset and expense accounts as decreases. Thus, the points d and c respectively mean an increase in monetary quantity ($d, c > 0$), and b and a mean a decrease in monetary quantity ($b, a < 0$). Although b and a are a negative number, this does not mean that the decrease in monetary quantity represented by them is intrinsically negative, as meant by the

[1] Setting up the debit and credit coordinate system as shown in the figure is a matter of convenience. The debit and credit labels may be interchanged. Only the interpretations differ.

nonnegativity-of-money convention in accounting (see Sect. 2.2, Chap. 2). Showing an increase or a decrease in individual accounts as a positive or negative number is a matter of presentation, not substance. What the lines Od, Oc, Ob, and Oa respectively show is the same as what the points d, c, b, and a respectively mean.

Given the specifications, the two-dimensional Cartesian space in the figure can serve as a basic geometric setting for presenting all types of accounting events according to the double entry principle.

Presenting Accounting Events: A Simple Case

If accounting events are of simple type, such as [d c] or [d a] in journal entry form, they have a single account (entry) on both debit and credit sides. Thus, they can be presented as a single point or line, or as a single (flat) space, in the two-dimensional (debit-credit) Cartesian space. This is what the shaded square in Fig. 3.1 shows. The square, and the circle therein, represents a type 1 accounting event that has a single account (entry) on both debit and credit sides of the journal entry. The accounting event is assumed to increase the debit and credit accounts alike (e.g., cash as a debit and revenue as a credit). Thus, $d, c > 0$ and the shaded square lies in QI of the two-dimensional space. It is a square, because it is defined by the pair of debit and credit accounts with the same monetary quantity ($d = c$) and the debit and credit coordinates are mutually perpendicular. Conceptually, the square is considered to reflect the duality and balance features of the accounting event and the contrasting yet symmetric nature of double entries.

In this book, a square and the circle therein such as the one in the figure are referred to as the double entry (DE) square and DE circle respectively of accounting events.[2] They are equivalent since they both represent the same accounting event and reflect the same duality and balance features of the event. The duality and balance features of the accounting event require the shaded space to be a square. The diagonal OD of the square has an angle of 45° with the adjacent debit and credit coordinates. As such, it has a slope of 1, equal to tangent 45° (tangent 45° = 1). The alternative geometric presentations—DE square and DE circle—of the accounting event are equivalent. Thus, if one of them is known or determined, the other is also known or determined. Similarly, if any of them is known or determined, the points and lines for the debit and credit elements of the accounting event are also known or determined.

The accounting event in Fig. 3.1 is of type 1 and assumed to increase the account on both debit and credit sides. Hence, the DE square and DE circle lie in QI of the debit-credit coordinate system. If an accounting event is of the same type but

[2] The DE square (DE circle) has different meanings in this book, depending on how it is defined or used. It is a geometric space representing accounting events in this chapter, a geometric presentation of the power of accounting events in Chap. 5, the double entry (DE) vector space of accounting events in Chap. 6 and a geometric space represented by the double entry (DE) matrix of accounting events in Chap. 7.

decreases the account on each side, they lie in QIII. If neither is the case, they lie in QII or QIV.

If an accounting event is of type 2, 3, or 4, thus having multiple accounts on the debit or credit side or on both sides, the event is considered a bundle of component transactions, which are executed in bundle as the accounting event is executed. As such, the accounting event is represented not by a single DE square but by multiple DE squares as a set. Whether or not these squares lie in the same quadrant of the two-dimensional space depends on whether or not the accounting event changes (increases or decreases) the multiple accounts on the debit and credit sides respectively in the same direction. If it does, those discrete DE squares lie in the same quadrant of the two-dimensional space. If it does not, they lie in different quadrants.

Since there is no third coordinate in the two-dimensional (debit-credit) space, the monetary size (volume) of an accounting event cannot be shown as a point or line on a distinct coordinate. Yet it can still be shown as a point or line, or even as a space. This point, and the line, lies *off* the debit and credit coordinates. And the size of the accounting event can be measured and shown by the size of the DE square or the DE circle therein. Since these alternative size measures are equivalent, they can be converted not only to each other but also to the same monetary size of the accounting event.

The diagonal OD of the DE square of type 1 accounting events divides the square into two contrasting yet symmetric right triangles of equal size as shown in the figure before. Thus, the space-based measure of the monetary size of such accounting events can be computed as

$$Q_S = d.c \quad (d = c)$$

or

$$\begin{aligned} Q_S &= (|OD| \text{ cosine } \theta).c \text{ or } d.(|OD| \text{ sine } \theta) \quad (\theta = 45^\circ) \\ &= (|OD| \text{ cosine } \theta)^2 \text{ or } (|OD| \text{ sine } \theta)^2 \\ &= |OD|^2 \text{ cosine}^2\, \theta \text{ or } |OD|^2 \text{ sine}^2\, \theta \end{aligned}$$

where |OD| denotes the length of the diagonal (line OD) of the DE square. Q_S is equal to the size of the DE square. It is a geometric measure of the (monetary) size of the accounting events. The larger the size of the events, the larger the size of the DE square of the events. Since d = c under the double entry principle, the product d.c. is equal to the size of the DE square. As $\theta = 45°$, cosine θ = sine θ. And, d = |Od| = |OD| cosine θ or c = |dD| = |OD| sine θ. Given this, the size of the DE square can also be computed according to the law of cosine or the law of sine in trigonometry as shown above.

Alternatively, the size of type 1 accounting events can also be measured by the size of the enclosed DE circle as

$$\pi r^2 = \pi(d/2)^2 \text{ or } \pi(c/2)^2$$

where r denotes the radius of the DE circle. The diameter of the DE circle is equal to a side of the DE square, which is in turn equal to the length of the line Od or Oc in the figure before. Although the DE circle differs in shape and size from the DE square, the two space-based measures of the size of the accounting events are equivalent. For a given accounting event, they can be converted not only to each other but also to the same monetary size of the event.

For an accounting event, the two space-based measures, DE square and DE circle, of the size of the event are equivalent. The ratio of the DE circle to the DE square is $\pi/4$ as shown below.

$$\pi(d/2)^2/Q_S = \pi(d/2)^2/d^2 = \pi/4 = 0.7854$$

or

$$= \pi(c/2)^2/c^2 = \pi/4 = 0.7854$$

Interestingly, this ratio is the same for all accounting events, regardless of how complex they are and how large their monetary size is. It is the same, regardless of whether accounting events are of type 1 or type 4 or whether their monetary size is large or small.

In theory, there can be an almost countless number of DE squares (or DE circles) in the entire two-dimensional (debit-credit) Cartesian space. Each of them represents a distinct accounting event. What this means is that an equally countless number of accounting events can be presented geometrically in the entire two-dimensional space. To see it, consider the DE square of an accounting event as a sheet of square paper placed on the debit-credit plane. It can be moved around from one location to another, without changing its size. The square paper at each location represents a unique accounting event. There are a countless number of such locations and a countless number of ways to move the square paper around from one location to another. Thus, there can be an equally countless number of accounting events that can be shown as a DE square in the entire two-dimensional space. If the size of the DE square at each location is changed as well, the total number of accounting events that can be shown as a DE square (DE circle therein) in the entire two-dimensional space rises to a level that can only be imagined.

Presenting Accounting Events: A Complex Case

More complex than the type 1 accounting event seen above are type 2, 3, and 4 accounting events. These accounting events have component transactions and multiple entries (accounts) on the debit or credit side (or both sides). The multiple accounts on each side may or may not change in the same direction. The component

Fig. 3.2 A two-dimensional (debit-credit) space for the presentation of complex-type accounting events

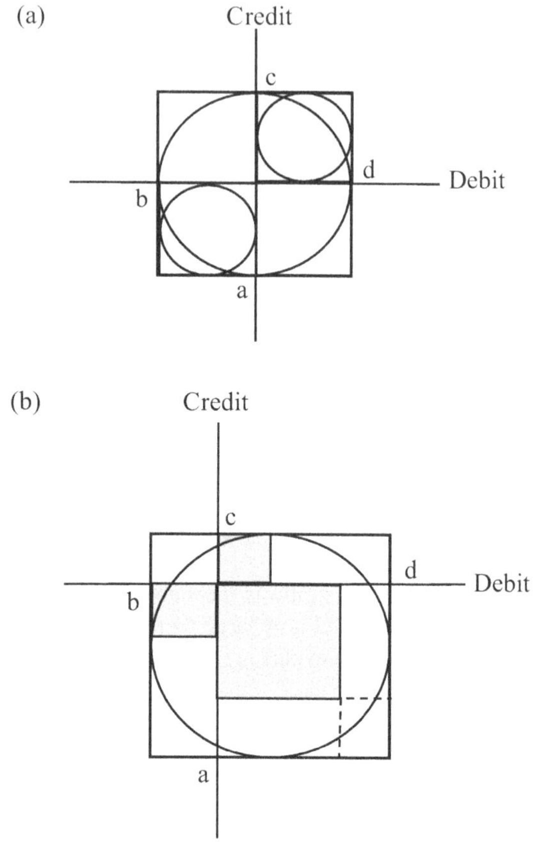

transactions are executed in bundle when the events are executed. As such, the monetary quantity of the debit entries and credit entries respectively of such accounting events is not one continuous quantity but a multitude of discrete quantities. So is their monetary size.

What the above means geometrically is that the DE square (DE circle) of such accounting events is not a single square (circle) but is divided into multiple squares (circles) or multiple squares and rectangles, depending on whether or not their component transactions have the same monetary size. And these squares and rectangles may or may not lie in the same quadrant of the two-dimensional (debit-credit) space since the multiple accounts on the debit or credit side (or both sides) may or may not change in the same direction. Thus, presenting such accounting events geometrically and measuring their size in the two-dimensional geometric space is somewhat involved. The size of accounting events is measured and shown by the size of those discrete squares and rectangles combined.

Figure 3.2 gives two examples of how to present complex accounting events in the debit-credit Cartesian space. One of the accounting events is shown in Fig. 3.2a. This event is assumed to be of type 4 and have two component transactions, one recorded as [d c] and the other as [b a] in journal entry form with b, a < 0. Thus, it is represented by the large unshaded DE square, which lies across the four quadrants of the two-dimensional space. The debit entry of the accounting event has a monetary quantity equal to d + |b| in absolute terms, and the credit entry has a monetary quantity equal to c + |a|. And the two component transactions are assumed to have the same monetary size. Thus, d = |b| and c = |a| for the component transactions respectively and d + |b| = c + |a| for the accounting event as a whole, under the double entry principle. And d.c. = d.|a| = |b|.c = |b|.|a| (in absolute terms), each of which is equal to the size of a component square, shaded or unshaded. The size of the DE square of the accounting event equals the product (d + |b|)(c + |a|). That is, it equals the size of the two shaded DE squares combined plus the size of the two unshaded squares combined. The component transaction denoted as [d c] is assumed to increase the debit and credit accounts alike. Thus, it is represented by the small shaded DE square (circle) in QI of the debit-credit coordinate system. The size of this DE square is equal to d.c. The other component transaction denoted as [b a] is assumed to decrease the debit and credit accounts alike. And it is represented by the small shaded DE square in QIII. The size of this DE square is equal to |b|.|a|. The size of the unshaded square in QII (QIV) is equal to the cross product |b|.c (d.|a|)—the product of the debit of one component transaction and the credit of the other.

Given the two component transactions, the DE square of the accounting event is made up of the four small squares of equal size and lies across the four quadrants. As a square, it reflects the duality and balance features of the accounting event. Similarly, each shaded component DE square also reflects the duality and balance features of the related component transaction. While the two unshaded squares are part of the DE square of the accounting event, neither one is a DE square since it does not represent any component transaction. In other words, the accounting event has no component transaction that is recorded as [b c] or [d a] in journal entry form.

When measured geometrically, the size of the accounting event equals the size of the large unshaded DE square or the DE circle therein. Each is computed as

$$Q_S = (d + |b|)(c + |a|) = d.c + d.|a| + |b|.c + |b|.|a|$$

or

$$Q_S = \pi r^2 = \pi[(|d| + |b|)/2]^2 \text{ or } \pi[(|c| + |a|)/2]^2$$

The direct product d.c. in the first expression above equals the size of the shaded DE square in QI, and |b|.|a| equals the size of the shaded DE square in QIII. The cross product |b|.c (d.|a|) equals the size of the unshaded square in QII (QIV). Since b and a are a negative number, they are shown in absolute value. This poses no problem since the size of a geometric space is an absolute (scalar) quantity and both b and a represent a monetary quantity, which is nonnegative in substance as meant by the nonnegativity-of-money convention (see Sect. 2.2, Chap. 2). Whether a debit or credit is shown as a negative or positive number is a matter of presentation, not substance.[3] The sign (+ or −) of the debit and credit of each component transaction determines the location, not size or shape, of the DE square of the component transaction.

For many complex accounting events, it is unlikely that their component transactions have the same monetary size. Nor are the multiple debit accounts or credit accounts likely to change in the same direction. For such accounting events, the following does not hold: $d = |b| = c = |a|$. In other words, the unshaded small squares in Fig. 3.2a are not squares but more likely to be rectangles. Fig. 3.2b presents one such accounting event. This event is assumed to be of type 4 and have three component transactions, each represented by a shaded small DE square. These component transactions differ not only in monetary size but also in the way they change the multiple accounts on each side, debit and credit. As such, the large unshaded DE square of the accounting event lies across the four quadrants QI to QIV. And the unshaded rectangles differ in size as well as location. As the accounting event has three component transactions, the debit entry of the event is divided into three discrete quantities (accounts)—d_1, d_2, and b. So is the credit entry—c, a_1, and a_2. For the accounting event as a whole, $d_1 + d_2 + |b| = c + |a_1| + |a_2|$ in monetary terms under the double entry principle. As it is, the unshaded DE square of the accounting event is divided into three small DE squares and six rectangles. The three DE squares respectively are defined by [d_1 c], [b a_1], and [d_2 a_2] in journal entry form, each representing a component transaction. Each rectangle is determined by the debit of one component transaction and the credit of another, or vice versa. Thus, the size of each rectangle is equal to the cross product of the debit (credit) of one component transaction and the credit (debit) of another (e.g., $d_2.c$ and $b.c$). As the monetary size of the component transactions differs, so does the size of the small DE squares representing them. There is no component DE square in QII since the accounting event has no component transaction that is recorded as [b c] in journal entry form.

If measured geometrically, the size of the accounting event in Fig. 3.2b equals the size of the three component DE squares combined plus the six rectangles combined, or the size of the large enclosed DE circle. Each is computed below.

$$
\begin{aligned}
Q_S &= (d_1 + d_2 + |b|)(c + |a_1| + |a_2|) \\
&= d_1.c + d_1.|a_1| + d_1.|a_2| + d_2.c + d_2.|a_1| + d_2.|a_2| + |b|.c \\
&\quad + |b|.|a_1| + |b|.|a_2|
\end{aligned}
$$

or

$$Q_S = \pi r^2 = \pi[(d_1 + d_2 + |b|)/2]^2 \text{ or } \pi[(c + |a_1| + |a_2|)/2]^2$$

[3] For the same reason, hereafter, all such negative numbers are considered or shown in absolute value when they mean a monetary quantity or a nondirectional scalar quantity such as distance, length, size, and space. This is consistent with the nature of quantity as a basic concept. Conceptually, a quantity itself is considered nonnegative in substance (see Sect. 2.2, Chap. 2).

There are three direct products of the debit and credit of the component transactions, all shown in *italics*. Each of them equals the size of a component DE square. And each of the six cross products of the debits and credits of the component transactions equals the size of a rectangle. The size of the accounting event measured by Q_S equals the size of the three component DE squares combined plus the size of the six rectangles combined. That is, it is equal to the size of the large unshaded DE square.

The graphical illustrations thus far suggest that the size (volume) of an accounting event can be measured and shown by either the size of the DE square of the event as one continuous quantity or the combined size of the component spaces (squares and rectangles) as a multitude of discrete quantities. Regardless, measuring the size of the accounting event as shown is consistent with the notion of quantity in general. Conceptually, a quantity of an object or a thing, whether tangible or intangible, can be either a magnitude or a multitude.[4] In other words, a quantity can be either one continuous magnitude or a multitude of discrete magnitudes, depending on how it exists. The monetary size of accounting events is one such quantity. As such, the size of accounting events can be one continuous magnitude or a multitude of discrete magnitudes, depending on whether the events are of type 1 or type 2, 3, or 4.

Conceptually, measuring the size of accounting events as a multitude of discrete magnitudes is consistent with the notion of linear addition or linear aggregation. It is also consistent with the well-known "whole-parts" relation. A linear aggregation means the addition of measures or quantities, each scaled by a factor of 1. It is like $A = \sum a_k$. In order for this to hold, the measures or quantities being added should be independent, meaning that those measures or quantities have no interaction with one another. Any interaction among them can make the aggregate A differ from the linear sum of the component measures or quantities. The "whole-parts" relation means that the "whole" is the same as the sum of its "parts," where the "whole" means one magnitude and the sum of the "parts" means a multitude of discrete magnitudes. For a given accounting event, the whole means the accounting event itself, and the parts mean its component transactions. And the size of the accounting event is the "whole" and the size of each component transaction is a "part." The "whole-parts" relation holds for complex accounting events as long as there is no interaction among the parts—the size measures of the component transactions.

As a quantity can be either one continuous magnitude or a multitude of discrete magnitudes, so can the monetary size of accounting events. The size of type 1 accounting events, which have no component transactions, is one continuous (monetary) magnitude. And the size of type 2, 3, or 4 accounting events, which have component transactions, is a multitude of discrete (monetary) magnitudes. Measuring the size of the latter-type accounting events by the size of their component transactions combined is consistent with the notion of linear aggregation and the "whole-parts" relation.

[4] See https://en.wikipedia.org/wiki/Quantity (July 2020).

Given the above, and consistent with the concept of quantity in general as a magnitude or a multitude, the space-based measure of the size of type 2, 3, or 4 accounting events having component transactions can be computed as

$$Q_S = \left(\sum |d_k|\right)\left(\sum |c_k|\right)$$

where d_k and c_k are the (monetary) quantity of the debit and credit entries (accounts) respectively of the kth component transaction of the accounting events. For a given accounting event, $\sum|d_k| = \sum|c_k|$ under the double entry principle. Since a decrease in an account in the debit and credit entries is shown as a negative number in the debit-credit coordinate system, all component quantities in the expression above are shown in absolute terms (for notational convenience). This does not pose a problem since the size of a geometric space, namely the size of accounting events in this case, is an absolute scalar quantity and a decrease in monetary quantity is nonnegative in substance. While the positive or negative sign (+ or −) of the component quantities affects the location of the DE square in the two-dimensional (debit-credit) space, it does not affect the size of the DE square.

As in the case of the type 1 (simple) accounting event before, the DE circle-to-DE square ratio for type 2, 3, or 4 accounting events is $\pi/4 = 0.7854$. If, for example, a type 4 accounting event has two component transactions and two entries (accounts) on both debit and credit sides, one entry as an increase and the other as a decrease on each side, the DE circle-to-DE square ratio for the accounting event is computed in general notations as

$$\begin{aligned} \pi r^2/Q_S &= \pi[(d + |b|)/2]^2/(d + |b|)^2 \\ &= \pi[(c + |a|)/2]^2/(c + |a|)^2 \\ &= \pi/4 = 0.7854 \end{aligned}$$

The double entry principle requires that $d + |b| = c + |a|$ in monetary quantity. The term $(d + |b|) / 2$, or $(c + |a|) / 2$, is equal to the radius of the DE circle. The DE circle-to-DE square ratio for the type 4 accounting event is $\pi/4$—the same as the ratio for the type 1 accounting event seen before. Again, this is because the ratio is not affected by the complexity or simplicity of accounting events. Nor is it affected by the size of accounting events. It is the same, regardless of whether accounting events are of type 1 or type 4 or whether their size is large or small. This is consistent with the notion of linear aggregation and the "whole-parts" relation indicated before.

3.2 Accounting Events in Two-Dimensional Space: Examples

How the two-dimensional (debit-credit) geometry of double entry accounting discussed thus far in this chapter works can be explained with various transaction examples. Using four different types of transaction examples, this section explains how to present them according to the double entry principle, and how to measure their (monetary) size, in the two-dimensional debit-credit Cartesian space. The four types of transaction examples are comparable to the four types of accounting events in journal entry form discussed in Sect. 2.4, Chap. 2. The illustrations in this section focus on the space-based presentation of accounting events and the space-based measurement of their size. The line- and point-based presentation of accounting events and the line- and point-based measurement of their size are illustrated later in this chapter.

Type 1 Events: A Single Account (Entry) on both Sides

Accounting events of type 1 are those that have a single account (entry) on both debit and credit sides of the journal entry, denoted as [d c] or [d a] for example. They have no component transactions. Suppose that a firm sells goods for $100 in cash. The journal entry to record this accounting event is.

Cash	$100	
Sales Revenue		$100

Figure 3.3 presents the accounting event in the two-dimensional (debit-credit) Cartesian space. The horizontal coordinate represents the debit dimension of the event, and the vertical coordinate represents the credit dimension. The coordinate for the size (monetary quantity) dimension of the event is suppressed as there is no third dimension for it in the two-dimensional space. The point 100 on the debit coordinate represents the cash $100 as an increase, and the point 100 on the credit coordinate represents the sales revenue $100 as an increase. The shaded square (circle) is the DE square (DE circle) of the accounting event. It has four equal sides, each being equal to 100. Since the event increases the debit and credit accounts alike, the DE square (DE circle) lies in QI of the two-dimensional space. It is a square as it reflects the duality and balance features of the accounting event and the contrasting yet symmetric nature of double entries.

The size of the accounting event, measured in monetary terms, equals $100 ($q = 100$), which is equal to a side of the shaded DE square. If measured geometrically, it equals the size of the DE square or the DE circle therein. Each is computed below.

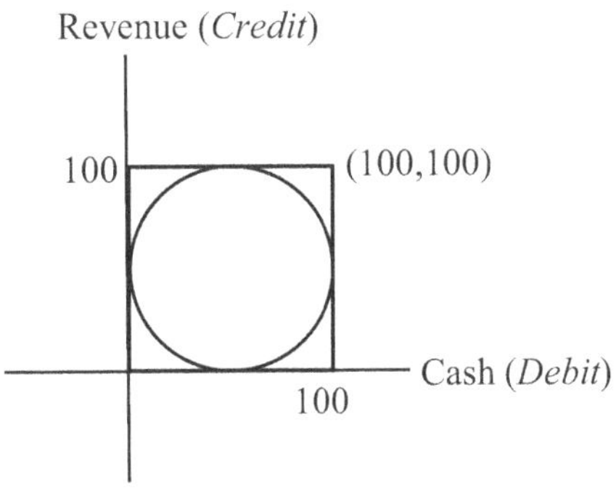

Fig. 3.3 Presenting a type 1 accounting event in the two-dimensional (debit-credit) space

$$Q_S = d.c = 100.100 = 100^2 au^2$$

or

$$Q_S = \pi r^2 = \pi(100/2)^2 = 7,854\ au^2$$

where au^2 means an "area unit" (e.g., square inch). The size of the accounting event can also be measured by the size of the DE circle shown since this circle represents the accounting event as well. Although different in magnitude, the two space-based measures of the size of the accounting event are equivalent. The DE circle-to-DE square ratio is $\pi r^2/Q_S = \pi/4 = 7854 / 100^2 = 0.7854$.

As the two space-based measures of the size of the accounting event are equivalent, they can be converted to the same monetary measure ($100):

$$q = (Q_S)^{1/2} = \left(100^2\right)^{1/2} = 100$$

or

$$q = 2r = 2[(Q_S)(0.7854)/\pi]^{1/2} = 100$$

The type 1 accounting event above increases the debit and credit accounts alike. Thus, the DE square (DE circle) lies in QI. If an accounting event is of the same type but decreases the account on each side, the DE square (DE circle) lies in QIII. If neither is the case, it lies in QII or QIV as the debit and credit accounts do not change in the same direction. Regardless, such accounting events can still be shown as a DE square or DE circle, and their size can be measured in the same way. And the DE circle-to-DE square ratio remains the same, 0.7854.

Type 2 Events: Multiple Accounts (Entries) on the Debit Side

Accounting events of this type have multiple accounts (entries) on the debit side. Suppose that a firm sells goods for $60 in cash and $40 on account. Given below is the journal entry to recognize the accounting event.

Cash	$60	
Account Receivable	40	
Sales Revenue		$100

As the event has two accounts on the debit side, it has two component transactions, which are executed in bundle as the accounting event is executed. One of the component transactions is the cash transaction, recorded as $[d_1\ \ c_1] = [60\ \ 60]$ in journal entry form, and the other is the account-receivable transaction, recorded as $[d_2\ \ c_2] = [40\ \ 40]$. Given the two component transactions, the debit and credit entries respectively of the event are made up of two discrete quantities. And the double entry principle requires $d_1 + d_2 = c_1 + c_2$.

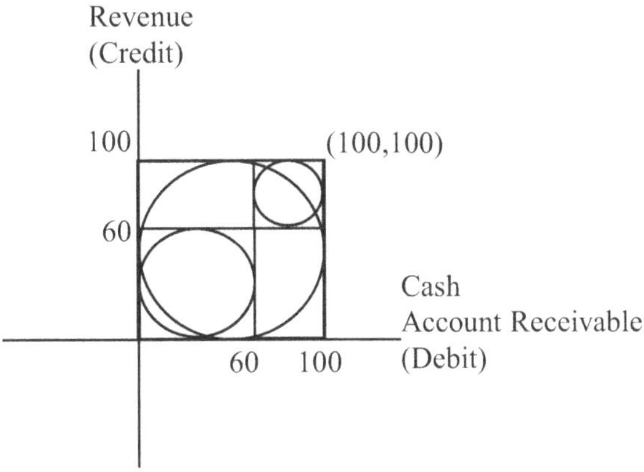

Fig. 3.4 Presenting a type 2 accounting event in the two-dimensional (debit-credit) space

Figure 3.4 presents the accounting event in the two-dimensional (debit-credit) Cartesian space. The large unshaded square (circle) is the DE square (DE circle) of the event. As the event increases the accounts on both sides of the journal entry, the DE square (DE circle) lies in QI. The accounting event has two component transactions so that the DE square of the event consists of two component DE squares and two rectangles as shown. The component square in the lower-left corner of the unshaded DE square represents the cash part of the accounting event, and the one in the upper-right corner represents the account-receivable part. Each component DE square has an enclosed circle of its own, which also represents the related component transaction. Besides the two component DE squares, there are two rectangles of the same size. One of them is determined by d_1 and c_2 and its size is equal to the cross product $d_1.c_2 = 60.40 = 2400$. And the other is determined by d_2 and c_1 and its size is equal to the cross product $d_2.c_1 = 40.60 = 2400$. Although divided into the discrete spaces, the DE square (DE circle) of the accounting event reflects the duality and balance features of the event and the contrasting yet symmetric nature of double entries.

If measured in monetary terms, the size of the accounting event equals $100 ($q = 100$). It is equal to a side of the unshaded DE square. If measured geometrically, it is equal to the size of the DE square or the DE circle therein. Each of these is computed below.

$$Q_S = (60 + 40)(60 + 40) = 100^2 \text{au}^2$$

or

$$Q_S = \pi r^2 = \pi((60 + 40)/2)^2 = 7,854 \text{ au}^2$$

Since the accounting event has two component transactions, Q_S is a multitude of discrete quantities. That is, the size measure Q_S equals the size of the two shaded component squares combined (60.60 + 40.40) plus the size of the two unshaded rectangles combined (60.40 + 40.60).

Measuring the size of the accounting event by the combined size of the discrete spaces does not affect the DE circle-to-DE square ratio: $\pi r^2/Q_S = \pi/4 = 0.7854$, which is the same as the ratio for the type 1 accounting event before. The ratio is not affected by the types of accounting events.

The two space-based measures above of the size of the accounting event are equivalent. Hence, they can be converted to the same monetary size ($100) of the event:

$$q = (Q_S)^{1/2} = \left(100^2\right)^{1/2} = \left(60^2\right)^{1/2} + \left(40^2\right)^{1/2} = 100$$

or

$$q = 2r = 2[(Q_S)(0.7854)/\pi]^{1/2} = 100$$

Unlike this event, a type 2 accounting event may increase one of the two debit accounts but decrease the other, while still increasing the single credit account. For example, a firm collects an interest revenue in cash and amortizes a bond premium, both as debit accounts, and recognizes the interest revenue as a single credit account. The DE square (DE circle) of this accounting event lies across QI and QII of the debit-credit coordinate system. As the event has two component transactions, it is represented by two component DE squares and two rectangles. One of the component DE squares lies in QI, and the other in QII. And one of the rectangles lies in QI, and the other in QII.

Type 3 Events: Multiple Accounts (Entries) on the Credit Side

This type of accounting events is the opposite of the type 2 accounting event seen above as they have a single debit account and multiple credit accounts. Suppose that a firm purchases goods for $100 by paying $60 in cash and holding $40 on account for future payment. The journal entry to record this event is.

Inventory	$100	
Cash		$60
Account Payable		40

This accounting event has two component transactions, one for the cash part of it and the other for the account-payable part. The cash part is recorded as $[d_1 \ \ a] = [60 \ \ -60]$ in journal entry form, and the account-payable part as $[d_2 \ \ c] = [40 \ \ 40]$. Unlike the previous accounting event, however, the cash part of this event decreases the cash account by $60.

Figure 3.5 graphically presents the accounting event in the two-dimensional debit-credit space. The large unshaded square (circle) is the DE square (DE circle) of the event. The goods purchased for $100 is shown as an increase in inventory assets on the positive side of the debit coordinate. The payment of $60 is shown as a decrease in cash on the negative side of the credit coordinate. And the unpaid amount $40 is shown as an increase in an account payable on the positive side of the credit

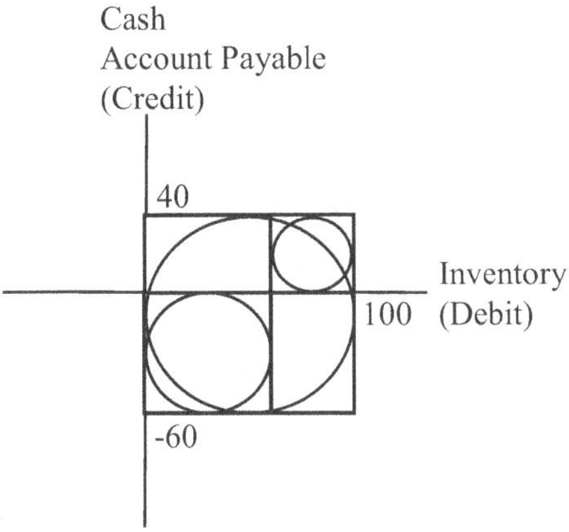

Fig. 3.5 Presenting a type 3 accounting event in the two-dimensional (debit-credit) space

coordinate. The cash paid is shown on the negative side of the credit, not debit, coordinate since it is credited to the cash account as a decrease in the journal entry. Given the above, the DE square (DE circle) of the accounting event lies across QI and QIV. As the event has two component transactions, the DE square is divided into two component DE squares and two rectangles. The small-shaded DE square in QIV (QI) represents the cash (account-payable) part of the accounting event. Each small square has a DE circle of its own, which also represents the related component transaction. As the two credit accounts do not change in the same direction, the two component DE squares do not lie in the same quadrant. Also, as the two component transactions do not have the same monetary size, the two component DE squares differ in size. Each of the two rectangles is determined by the debit of one component transaction and the credit of the other. They have the same size but do not lie in the same quadrant.

The size of the accounting event, measured in monetary terms, is \$100 ($q = 100$), which is equal to a side of the unshaded DE square of the event. It is equal to the distance between the two points 0 and 100 on the debit coordinate or the distance between the two points 40 and -60 on the credit coordinate. If measured in the geometric space, the size of the accounting event equals the size of the large DE square or the size of the DE circle therein. Each is computed as

$$\begin{aligned} Q_S &= (60 + 40)(40 + |-60|) \\ &= 60.40 + 60.|-60| + 40.40 + 40.|-60| \\ &= 100^2 \text{au}^2 \end{aligned}$$

or

$$\begin{aligned} Q_S &= \pi r^2 = \pi[(60 + 40)/2]^2 \\ &= 7,854 \text{ au}^2 \end{aligned}$$

The size measure Q_S in the first expression equals the size of the two component DE squares combined plus the size of the two rectangles combined. The direct product $60.|-60|$ equals the size of the DE square (in QIV) of the cash part of the accounting event, and 40.40 equals the size of the DE square (in QI) of the account-payable part. The cross product 60.40 is equal to the size of the rectangle in QI, and $40.|-60|$ is equal to the size of the rectangle in QIV. The size of the accounting event can also be measured by the size of the DE circle as shown. In computing Q_S, the negative number -60 for the decrease in cash is shown in absolute value.[5] Showing the decrease in cash as a negative number does not pose a problem since the size of a geometric space is an absolute scalar quantity. The positive or negative sign (+ or −) of the component quantities affect the location, not size, of the DE square (DE circle).

As seen, the size of the accounting event measured by Q_S is not one continuous magnitude but a multitude of discrete magnitudes. Again, measuring the size of the accounting event as such a multitude is reasonable for the reason indicated before. It is consistent with the notion that the "whole" is the same as the sum of its "parts" and that aggregation is a linearly additive concept. As in the case of the previous accounting events, the DE circle-to-DE square ratio for this accounting event is $\pi r^2/Q_S = \pi/4 = 0.7854$.

Since the two space-based measures of the size of the accounting event are equivalent, they can be converted to the same monetary size ($100) of the event as shown below.

$$q = (Q_S)^{1/2} = \left(100^2\right)^{1/2} = 100$$

or

$$q = 2r = 2[(Q_S)(0.7854/\pi)]^{1/2} = 100$$

Unlike the type 3 transaction above, a type 3 accounting event may increase multiple credit accounts as well as a single debit account. For example, a firm purchases an inventory asset on account with a purchase discount taken. If so, the DE square and component DE squares, and the related rectangles, of the accounting event all lie in QI of the debit-credit coordinate system. If a type 3 accounting event decreases the single debit account and one of the two credit accounts (e.g., the payment of an account payable in cash with a purchase discount taken), one of the component DE squares lies in QII, and the other in QIII. And the DE square of the accounting event lies across QII and QIII. Yet the size of the DE square and the DE circle respectively is still computed in the same way. And the DE circle-to-DE square ratio remains the same: $\pi r^2/Q_S = \pi/4 = 0.7854$.

[5] As indicated in a footnote before in this chapter, hereafter, all such negative numbers are considered in absolute value when nondirectional scalar quantities such as distance, length, and space are computed.

Type 4 Events: Multiple Accounts (Entries) on Both Sides Accounting events of this type are more complex than those seen before since they have multiple accounts (entries) on both debit and credit sides of the journal entry and the multiple entries on each side can have multiple accounts as a mix of increases and decreases. Suppose that a firm recognizes a current-period interest expense of $330 on a bond liability, which was issued at a discount some years ago, and pays an accrued interest of $350 in cash on the bond liability. The interest expense recognized includes a discount amortization of $100. And the interest $350 paid in cash includes a previously accrued but unpaid interest of $120 and a newly accrued interest of $230 for the current period.[6] Given below is the journal entry made to record the transaction.

Interest Payable	$120	
Interest Expense	330	
Cash		$350
Bond Payable (discount amortization)		100

The accounting event consists of three component transactions, which are executed in bundle as the transaction is executed. Given below are the journal entries to record the component transactions.

Interest Payable	$120	
Cash		$120
Interest Expense	$230	
Cash		$230
Interest Expense	$100	
Bond Payable (discount amortization)		$100

The three component transactions above can also be shown in an abstract journal entry form [debit credit] as

$[b \;\; a_1] = [-120 \;\; -120]$ for the past-period interest paid
$[d_2 \;\; a_2] = [230 \;\; -230]$ for the current-period interest paid
$[d_1 \;\; c] = [100 \;\; 100]$ for the bond discount amortized

And the double entry principle requires $|d_1 + d_2| + |b| = |c| + |a_1 + a_2|$.

Figure 3.6 presents the accounting event in the two-dimensional (debit-credit) Cartesian space. The large unshaded square and the circle therein are the DE square and DE circle respectively of the event. It consists of three small DE squares and six rectangles as the accounting event has three component transactions. Each of these six rectangles is determined jointly by the debit (credit) of one component

[6]It is assumed that the interest payment date of the bond liability does not match the fiscal-year end of the firm and that the unpaid interest of $120 was recognized as an interest payable at the end of the past fiscal year, which is paid in cash today—the interest payment date for the bond liability.

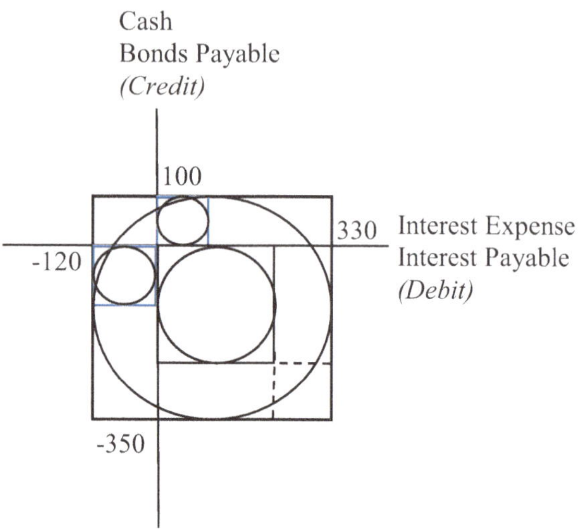

Fig. 3.6 Presenting a type 4 accounting event in the two-dimensional (debit-credit) space

transaction and the credit (debit) of another. The large DE square (DE circle) reflects the duality and balance features of the accounting event and the contrasting yet symmetric nature of double entries. Each small DE square represents a component transaction and has an enclosed DE circle. Like the large DE square of the accounting event, it also reflects the duality and balance features of the related component transaction. The accounting event has two accounts on each side of the journal entry, one as an increase and the other as a decrease. Thus, the DE square lies across four quadrants of the debit-credit coordinate system, with vertices: (330,100) in QI, (−120,100) in QII, (−120,−350) in QIII, and (330,−350) in QIV. The interest \$120 paid in cash is shown on the negative side of the debit coordinate since it is debited to the interest payable account as a decrease in the journal entry. Consistently, the interest \$350 paid in cash is shown on the negative side of the credit coordinate since it is credited to the cash account as a decrease in the journal entry.

The size of the accounting event, measured in monetary terms, is \$450 ($q = 450$), which is equal to a side of the large DE square. It is equivalent to the distance between the two points 330 and −120 on the debit coordinate or between the two points 100 and −350 on the credit coordinate. If measured in the geometric space, the size of the accounting event is equal to the size of the large DE square or the DE circle therein, computed as

$$\begin{aligned} Q_S &= (100 + 230 + |-120|)(100 + |-120| + |-230|) \\ &= 202,500 \text{ au}^2 \end{aligned}$$

or

$$\begin{aligned} Q_S &= \pi r^2 = \pi(450/2)^2 \\ &= 159,043.1281\ au^2 \end{aligned}$$

The size measure Q_S in the first expression equals the size of the large DE square, which equals the size of the three shaded component DE squares combined plus the size of the six unshaded rectangles combined. The size of each small DE square is equal to the direct product of the debit and credit of the related component transaction. And the size of each rectangle is equal to the cross product of the debit (credit) of one component transaction and the credit (debit) of another.[7] As seen, Q_S is not one continuous quantity but a multitude of discrete quantities since the event has component transactions. Again, measuring the size of the accounting event as such a multitude poses no problem for the reason indicated before.

The complexity and size of this accounting event do not affect the DE circle-to-DE square ratio: $\pi r^2/Q_S = \pi/4 = 159{,}043.1281 / 202{,}500 = 0.7854$. It is the same as the ratio for the three previous transaction examples even if this event is more complex in nature and much larger in monetary size. Again, this is because the DE circle-to-DE square ratio is not affected by the complexity or size of accounting events.

The two space-based measures (square and circle) Q_S of the size of the accounting event are equivalent. Hence, they can be converted to the same monetary size of the event:

$$q = (Q_S)^{1/2} = 202,500^{1/2} = 450$$

or

$$q = 2r = 2[(Q_S)(0.7854)/\pi]^{1/2} = 450$$

which is also equal to

$$\begin{aligned} q &= \left(100^2\right)^{1/2} + \left(|-120|^2\right)^{1/2} + (230.|-230|)^{1/2} \\ &= 450 \end{aligned}$$

or

$$\begin{aligned} q &= 2\left(\pi 50^2/\pi\right)^{1/2} + 2\left(\pi 60^2/\pi\right)^{1/2} + 2\left(\pi 115^2/\pi\right)^{1/2} \\ &= 450 \end{aligned}$$

[7] To see all this algebraically, the expression of Q_S can be expanded as 100.100 + 100.|−120| + 100.|−230| + 230.100 + 230.|−120| + 230.|−230| + |−120|.100 + |−120|.|−120| + |−120|.|−230| = 202,500. Here, there are three direct products 100.100, 230.|−230|, and |−120|.|−120|, each being equal to the size of the related small DE square. And there are six cross products, each being equal to the size of the related rectangle.

where 50, 60, and 115 are the radius of the DE circles of the three component transactions.

3.3 Features of the Double Entry Square

The DE square (DE circle) of accounting events seen in this chapter has interesting features. First of all, it is a square (area) as it reflects the duality (dual sides) and balance features of accounting events and the contrasting yet debit-credit symmetric nature of double entries. As such, the diagonal of the DE square has an angle of 45° with the debit and credit coordinates and has a slope of 1 (tangent 45° = 1). The DE square has its two sides along, or in parallel with, the debit and credit coordinates, which are mutually orthogonal. The double entry principle requires it not to tilt toward or away from any coordinate. The DE square has a vertex at the point of origin (zero quantity) or the point of origin as an inner point. Or it has the point of origin on one of its sides. In other words, the DE square of accounting events cannot lie away from the point of origin. It can lie in any one quadrant of the two-dimensional (debit-credit) space or across any pair of adjacent quadrants—(QI, QII), (QII, QIII), (QIII, QIV), and (QI, QIV)—or across all four quadrants (QI, QII, QIII, QIV). However, it cannot lie across three quadrants because a square cannot be defined across three quadrants, while still having the features of the DE square described above.

Having such features, the DE square is not just any square but a special kind of square in the two-dimensional (debit-credit) space. As it is, not all squares (and inner circles therein) or similar geometric spaces can be a DE square (circle). Figure 3.7 presents some geometric shapes that look like a DE square (DE circle) but are actually not. None of them has the features of a DE square. Nor can any of the

Fig. 3.7 Examples of geometric shapes that are not a DE square (DE circle)

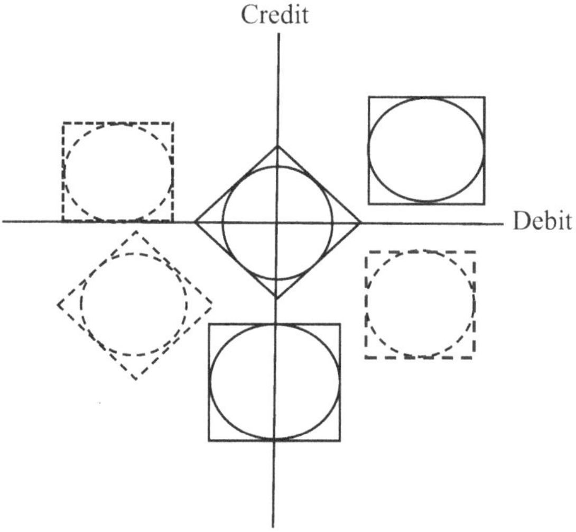

enclosed circles be a DE circle for the same reason. For example, the solid-line rhombus in the middle of the figure has the point of origin as an inner point and its vertices lie on the debit and credit coordinates. But none of its sides runs along or in parallel with the debit or credit coordinate. Nor does the diagonal of the rhombus have an angle of 45° with the adjacent coordinates. For similar and other reasons, neither the dashed-line rhombus in QIII nor the dashed-line square in QIV is a DE square. The latter has its sides in parallel with the debit and credit coordinates but lies away from the point of origin. For similar reasons, neither the solid-line square in QI nor the solid-line square lying across QIII and QIV is a DE square. The dashed-line square in QII has its two sides in parallel with the debit and credit coordinates but lies away from the point of origin. As none of the squares and rhombuses shown is a DE square, none of the circles therein is a DE circle.

3.4 Accounting Events as Lines and Points

As accounting events can be shown by geometric spaces, so can they be shown by lines and points. In geometry, a point is defined as a location in a geometric space. It has no dimension, length, magnitude (size), or shape. It is the most fundamental and primitive concept of all in geometry; other geometric concepts such as line and space are defined based on the concept of point. A line is the shortest distance, or an infinite number of collinear points, between two points. In a multidimensional Cartesian geometric space with a point of origin, a point has an angle from the point of origin. That is, it is a location with a direction or a directed location. It is shown by not just an array but ordered array of real numbers. Similarly, a line is a directed line and, like a point, is an ordered array of real numbers in a multidimensional Cartesian space.

In the two-dimensional (debit-credit) Cartesian space with the point of origin, the line and point for the debit and credit elements respectively of accounting events are an ordered array of two numbers, one of which is the point of origin (zero monetary quantity). And the line and point for each accounting event is also an ordered array of two numbers, one for the debit element of the event and the other for the credit element. Hereafter, this line is referred to as the double entry (DE) line of accounting events, and this point as the double entry (DE) point of accounting events. The former is the diagonal of the DE square of accounting events, and the latter is the diagonal vertex of the DE square. Each is determined jointly by the elements d and c of accounting events. Unlike the line and point for d and c, the DE line and DE point of accounting events lie *off*, not on, the coordinates of the two-dimensional space. So do the line and point representing the size element q of accounting events. The size of accounting accounting events can be measured and shown by the length of the DE line or the distance of the DE point from the point of origin. This is because there is no third coordinate for q in the two-dimensional space. The length of the DE line, and the distance of the DE point, can be converted to the monetary size of the accounting events. Discussed below are the line- and point-based presentation of accounting events and the line- and point-based measurement of their size using two transaction examples from Sect. 3.2.

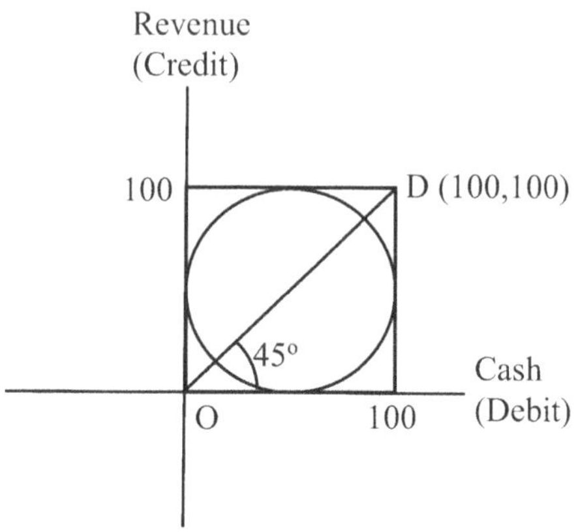

Fig. 3.8 The line- and point-based presentation of a type 1 accounting event in the two-dimensional (debit-credit) space

Example 1 Consider the type 1 transaction example (before)—that a firm sells goods for \$100 in cash. This transaction has a single account on both debit and credit sides. Figure 3.8, adapted from Fig. 3.3, presents the accounting event as a DE line and a DE point respectively in the two-dimensional (debit-credit) Cartesian space. The shaded square is the DE square of the accounting event. If the event is shown by a line, it is shown by the line OD—the DE line of the event. Like the DE square, this line reflects the duality and balance features of the accounting event and the contrasting and symmetric nature of double entries. It has a slope of 1 and an angle of 45° with the adjacent debit and credit coordinates (tangent 45° = 1). If the accounting event is shown by a point, it is shown by the diagonal vertex (point D) of the DE square. This point is the DE point of the event. Like the DE line of the event, it also has an angle of 45° from the point of origin. It reflects the same duality and balance features of the accounting event and the contrasting and symmetric nature of double entries. Since the accounting event increases cash as a debit and revenue as a credit, the DE line and DE point lie in QI of the debit-credit Cartesian space as the shaded DE square does.

Given the DE line and DE point, the size of the accounting event is measured by the length of the line OD or the distance of the DE point D from the point of origin. As the DE line has an angle of 45° with the debit and credit coordinates, it divides the DE square into two symmetric right triangles of equal size. Given this, the line-based measure (Q_L) of the size of the accounting event is computed according to the well-known Pythagorean theorem in geometry, or the law of cosine or the law of sine in trigonometry, as follows:

$$Q_L = \left(d^2 + c^2\right)^{1/2} = \left(100^2 + 100^2\right)^{1/2} = 141.4214$$

or

$$\begin{aligned} Q_L &= d/\text{cosine}\ \theta \text{ or } c/\text{sine}\ \theta \\ &= 100/\text{cosine}\ \theta \text{ or } 100/\text{sine}\ \theta \\ &= 100/0.70710678 = 141.4214 \end{aligned}$$

where d (c) is the debit (credit) element of the accounting event (d, c = 100). Q_L is equal to the length of the DE line of the event and to the distance of the DE point D from the point of origin.

The line-based measure Q_L of the size of the accounting event is equivalent to the space-based measure (Q_S) of the size of the event in the previous section because they both are a measure of the size of the same accounting event. Hence, the following holds:

$$\begin{aligned} Q_S &= d.c = (Q_L \text{cosine } \theta)(Q_L \text{sine } \theta) \\ &= (141.42136^2)(0.70710678^2) = 100^2 \text{au}^2 \end{aligned}$$

or

$$Q_S = d.|OD| \text{ sine } \theta \text{ or } c.|OD| \text{ cosine } \theta = 100^2 \text{au}^2$$

where |OD| is the length of the DE line OD. The expressions above show how the line-based measure Q_L of the size of the accounting event is related to the space-based measure Q_S of it.

Since the alternative line-based measures of the size of the accounting event are equivalent, they can be converted to the same monetary size of the event:

$$\begin{aligned} q &= d = |OD| \text{ cosine } \theta \text{ or } c = |OD| \text{ sine } \theta \\ &= Q_L \text{cosine } \theta \text{ or } Q_L \text{sine } \theta = (141.4214)(0.7071) \\ &= 100 \end{aligned}$$

or

$$q = Q_L/\sqrt{2} = 100$$

In general, the line- or point-based measure of the size of type 1 accounting events having a single account on both debit and credit sides is computed as

$$Q_L = (d^2 + c^2)^{1/2} = (2d^2)^{1/2} = d\sqrt{2} \text{ or } (2c^2)^{1/2} = c\sqrt{2}$$

or

$$Q_L = d/\text{cosine } \theta \text{ or } c/\text{sine } \theta \quad (\theta = 45^\circ)$$

Since Q_L is equivalent to the space-based measure Q_S of the size of the accounting events, the following hold:

$$\begin{aligned} Q_S &= d.c = (Q_L \text{ cosine } \theta)(Q_L \text{ sine } \theta) \quad (\theta = 45^\circ) \\ &= (Q_L)^2 \text{ cosine}^2 \theta \text{ or } (Q_L)^2 \text{sine}^2 \theta \end{aligned}$$

and

$$Q_S = d.|OD| \text{ sine } \theta \text{ or } c.|OD| \text{ cosine } \theta$$

And the line-based (nonmonetary) measure Q_L of the size of the accounting events can be converted to their monetary size as follows:

$$Q_L/\sqrt{2} = q$$

or

$$Q_L \text{ sine } \theta \text{ or } Q_L \text{cosine } \theta = q$$

Example 2 More complex than the previous accounting event is the type 4 transaction example (before)—that a firm recognizes an interest expense of $330, pays an interest of $350, and amortizes a discount of $100, all relating to a bond liability that was issued at a discount some time ago. Figure 3.9 graphically presents the accounting event in the two-dimensional (debit-credit) space. The event has three component transactions and two accounts on the debit and credit sides respectively one as an increase and the other as a decrease. Thus, the DE square of the accounting event lies across four quadrants of the two-dimensional space. As the event has three component transactions, it is represented by three DE lines, and three DE points, as a set. And there are three small DE squares, one for each component transaction. The small shaded DE square in QI of the two-dimensional space represents an increase in the bond liability by $100 due to the amortization of bond discount. The small DE square in QIII represents a decrease in cash by $120 due to the payment of the unpaid past-period interest $120. And the small DE cube in QIV represents an increase in the interest expense by $230 due to the accrual and payment of the current-period interest expense. The diagonal of each small DE square is a DE line, and the diagonal vertex is a DE point.

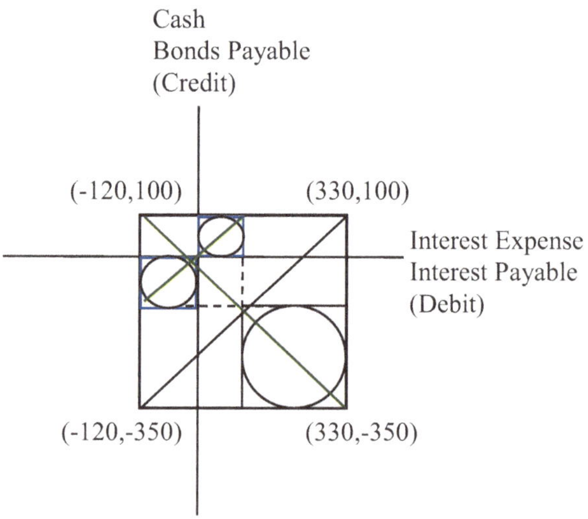

Fig. 3.9 The line- and point-based presentation of a type 4 accounting event in the two-dimensional (debit-credit) space

The length of the three DE lines combined is equal to the length of the diagonal of the large unshaded DE square of the accounting event. And the size of the event is measured by the length of the three DE lines combined or the distance of the three DE points combined. Although the three DE lines lie in different quadrants of the two-dimensional space, their lengths can be added since the length of lines in general is an absolute scalar quantity. The same is also true for the three DE points since the distance of points in general is also an absolute scalar quantity.

The DE line of each component transaction divides the related small DE square into two right triangles of equal size. Given this, the line-based measure of the size of the accounting event can be computed according to the Pythagorean theorem and the law of cosine respectively as follows:

$$\begin{aligned} Q_L &= \left(100^2 + 100^2\right)^{1/2} + \left(|-120|^2 + |-120|^2\right)^{1/2} + \left(230^2 + |-230|^2\right)^{1/2} \\ &= 636.3961 \end{aligned}$$

or

$$\begin{aligned} Q_L &= 100/\text{cosine}\ \theta + |-120|/\text{cosine}\ \theta + 230/\text{cosine}\ \theta \\ &= 450/0.7071 = 636.3959 \end{aligned}$$

where $\theta = 45°$. The two alternative measures Q_L of the size of the accounting event are equivalent as they both are a geometric measure of the size of the same event. Also, Q_L is equal to the combined distance of the three DE points of the component transactions from the point of origin. The line-based size measure Q_L of the event is computed as a multitude of discrete lines combined. Computing the size of the accounting event as such a multitude is consistent with the notion of linear aggregation and the "whole-parts" relation—that the "whole" is the same as the sum of its "parts."

The size measure Q_L is equivalent to the space-based measure Q_S of the size of the same accounting event in Sect. 3.2. Hence, the following holds:

$$\begin{aligned} Q_S &= d.c = (Q_L\text{cosine}\ \theta)(Q_L\text{sine}\ \theta) \\ &= \left(636.3961^2\right)(0.707107)^2 \\ &= 450^2\text{au}^2 \end{aligned}$$

where $d = 450 = Q_L$ cosine 45°, and $c = 450 = Q_L$ sine 45°, respectively.

The line-based measure Q_L of the size of the accounting event can be converted to the monetary size of the event as follows:

$$q = d \text{ or } c = Q_L/\sqrt{2} = 636.3961/\sqrt{2} = 450$$

or

Fig. 3.10 An alternative line- and point-based presentation of a type 4 accounting event in the two-dimensional (debit-credit) space

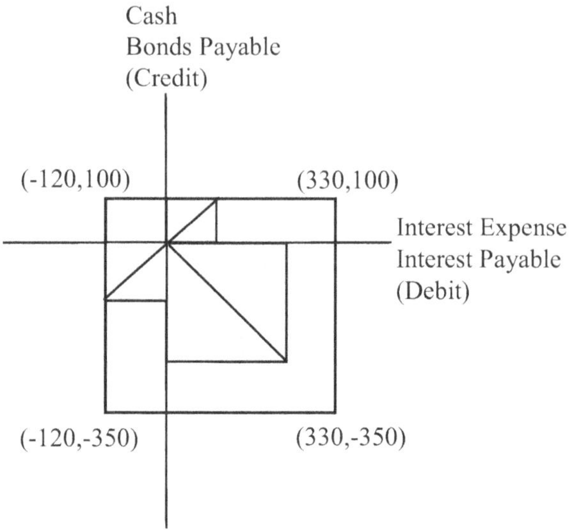

$$\begin{aligned} q &= Q_L \text{cosine } \theta \text{ or } Q_L \text{sine } \theta \\ &= (636.3961)(0.7071) = 450 \end{aligned}$$

Presenting the type 4 accounting event as shown in Fig. 3.9 is one possible way. Another way is to move one component square and the related DE line as shown in Fig. 3.10, similar to Fig. 3.6. The size of the accounting event is then measured by the combined length of the diagonals of the three shaded DE squares as follows:

$$\begin{aligned} Q_L &= D_1 + D_3 + D_4 \\ &= \left(100^2 + 100^2\right)^{1/2} + \left(|-120|^2 + |-120|^2\right)^{1/2} + \left(230^2 + |-230|^2\right)^{1/2} \\ &= 636.3961 \end{aligned}$$

where D_1, D_3, and D_4 denote the DE line of the three respective component transactions. The subscripts 1, 3, and 4 indicate the quadrants QI, QIII, and QIV respectively where those component DE lines lie.

In general, for type 2, 3, or 4 accounting events, the line-based measure Q_L of their size is computed as

$$Q_L = \left[\left(\sum |d_k|\right)^2 + \left(\sum |c_k|\right)^2\right]^{1/2} = \sum |d_k|\sqrt{2} \text{ or } \sum |c_k|\sqrt{2}$$

where d_k and c_k are the debit and credit elements respectively of the kth component transaction of the accounting events. And $|d_k| = |c_k|$ for each component transaction, and $\sum|d_k| = \sum|c_k|$ for the accounting event, under the double entry principle. The component quantities are shown in absolute value since some of them can be a decrease in monetary quantity, thus shown as a negative number on the debit and credit coordinates. Given this, and for notational convenience, all component quantities of d and c are shown in absolute terms. This poses no problem since the length of a line or the distance of a point, like the size of a geometric space, is an absolute scalar quantity.

The line-based measure Q_L above of the size of accounting events having component transactions can be converted to the monetary size of the events. This is shown in general notations below:

$$Q_L/\sqrt{2} = q \text{ or } \sum |d_k| \text{ or } \sum |c_k| = q$$

or

$$Q_L \text{cosine } \theta \text{ or } Q_L \text{sine } \theta = q \quad (\theta = 45^o)$$

3.5 Relative Merits of the Two-Dimensional Approach

When compared to the one-dimensional setting in Chap. 2, the two-dimensional debit-credit geometric setting in this chapter has some merits. Conceptually, it fits better for the multidimensional nature of accounting events. Also, it is more consistent with the feature of the double entry system as a four-way classification method. In the two-dimensional setting, not only can the debit and credit elements respectively of accounting events be recognized on a distinct dimension, but their increase and decrease can also be made explicit on the different sides, positive and negative, of the debit and credit coordinates, respectively. This is not possible in the one-dimensional (debit-credit) setting as seen in Chap. 2. Those merits of the two-dimensional setting are particularly pronounced when accounting events of type 2, 3, or 4 are presented geometrically. Accounting events of such types are complex as they have component transactions and the multiple accounts on the debit or credit side (or both sides) may or may not change in the same direction. Yet it is still possible to present such accounting events more clearly and have better (visual) insights into their complexity and details in the two-dimensional setting.

The relative merits of the two-dimensional (debit-credit) setting may be illustrated with specific transaction examples. Given below are the journal entries for three different transaction examples of the same monetary size. The journal entry (a) is made to record the sale of goods for \$80 in cash and \$20 on account, (b) to record the payment of an account payable with a purchase discount of \$20 taken, and (c) to record the payment of accrued interest and the amortization of a premium relating to a bond liability that was issued at a premium before.

(a)	Cash	$80	
	Account Receivable	20	
	Sales Revenue		$100
(b)	Account Payable	$100	
	Cash		$80
	Purchase Discount		20
(c)	Interest Expense	$80	
	Bond Premium (amortization)	20	
	Cash		$100

The three transactions have the same monetary size ($100) so that they can be shown as the same array of three numbers: (d, c, q) = (100, 100, 100). Although equal in monetary size, the three transactions differ not only in type but also in the debit and credit entries. Transactions (a) and (c) are of type 2, and transaction (b) is of type 3. Also, the former have two accounts on the debit side, while the latter has two accounts on the credit side.

In the one-dimensional (debit-credit) geometric setting, despite their differences, the three transactions are all shown by the same points or lines, d = 100 and c = 100, on the same single debit-credit dimension. In particular, transactions (a) and (c) are shown by the same points or lines even though they are two very different transactions. The points (lines) for the debit and credit entries of transaction (a) are the same as the points (lines) for the debit and credit entries of transaction (c). This is so, even if the two debit entries of the latter transaction are not the same kind of change—one being an increase and the other being a decrease. Also similar in this regard is the decrease in cash of $100 versus the increase in revenue of $100. Both of these are shown by the same point (line) on the same credit side of the single debit-credit dimension even if they are not the same kind of change. Further, the point for the monetary size of each transaction is not (cannot be) shown on the debit-credit single dimension. Similar difficulties also exist with transaction (b) versus transaction (c). Both the decrease in the account payable $100 and the increase in the interest expense $80 are shown on the same debit side of the single debit-credit dimension even if the two are not the same kind of changes. Also, similar in this regard are the increase in the purchase discount $20 and the decrease in the cash $100.

The difficulties with the one-dimensional geometry are greatly mitigated in a two-dimensional setting. Figure 3.11 presents the three transactions in the two-dimensional (debit-credit) space. Each shaded DE square represents a distinct transaction and distinguishes itself from the others. For transactions (b) and (c), the decrease in cash is shown on the negative (−) side of the credit, not debit, coordinate as it is credited to the cash account in the journal entry. And the decrease in both the account payable and the bond premium is shown on the negative (−) side of the debit, not credit, coordinate as it is debited to the respective account in the journal entry. The bond premium amortization of $20 as a decrease in the bond liability is shown on the negative side of the debit, not credit, coordinate. As the three

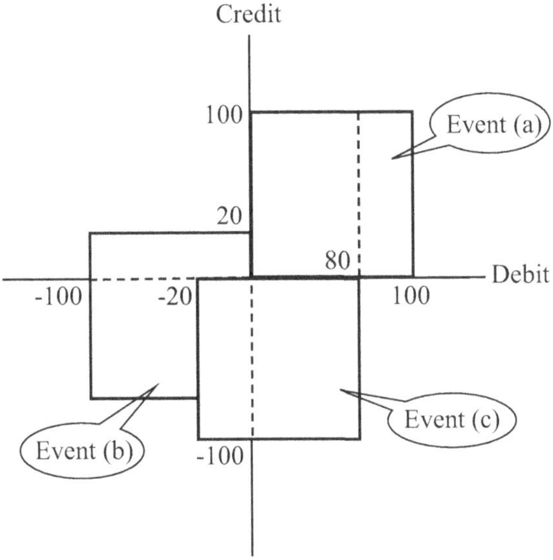

Fig. 3.11 Presenting different accounting events of the same monetary size in the two-dimensional (debit-credit) space

transactions have the same (monetary) size, the three shaded DE squares shown have the same size. Yet their locations clearly differ, indicating that they are three very different transactions. The distinct location of each DE square makes it easy to see not only how each transaction differs from the others but also how their component transactions differ from one another. What all this means is that the two-dimensional geometric setting, relative to the one-dimensional one, offers better (visual) insights into the unique nature of each transaction as well as their differences.

The Double Entry Accounting in Three-Dimensional Setting

4

This chapter extends the discussion of the geometry of double entry accounting in Chap. 3 to a three-dimensional geometric setting. It examines how to present accounting events and explores the features of double entry accounting in a three-dimensional space. It does so using the same four transaction examples in the previous chapter. The geometric setting used to do so is a debit-credit-size Cartesian space with three coordinates. In this space, each element (debit, credit, and size) of accounting events can be shown on a distinct dimension (coordinate), which was not possible in the two-dimensional (debit-credit) setting in Chap. 3. As such, the three-dimensional setting fits conceptually better for the multidimensional nature of accounting events. And it provides better (visual) insights into not only the features of accounting events but also the features of double entry accounting in general.

4.1 The Three-Dimensional Geometry of Double Entry Accounting

The Basic Setting

The geometric setting in this chapter is a three-dimensional Cartesian space with three coordinates, one for each dimension of accounting events—debit, credit, and monetary size. These dimensions are denoted as D, C, and S, and each is represented by the three elements d, c, and q respectively of accounting events. The third coordinate represents the size dimension S, and q is a point on this coordinate. The debit and credit elements d and c of accounting events are a point on the debit and credit coordinates respectively as they were in the two-dimensional space in Chap. 3.

Figure 4.1 gives an example of the three-dimensional (debit-credit-size) setting where accounting events can be presented geometrically. It is equivalent to Fig. 3.1. The three coordinates are set up not by following a convention in geometry but by

© The Author(s), under exclusive license to Springer Nature Switzerland AG 2025
B. T. Ro, *The Geometry of Accounting*, Springer Texts in Business and Economics,
https://doi.org/10.1007/978-3-031-88892-2_4

preserving the debit-credit coordinate system in Chap. 3.[1] The point O is the point of origin (zero monetary quantity). While the debit and credit coordinates have the negative (−) section, the size coordinate does not since the monetary size (volume) of accounting events cannot be negative. The size of accounting events is always positive ($q > 0$) and cannot even be zero ($q \neq 0$) since no accounting event can take place with a zero monetary volume. The three coordinates, along with the point of origin, divide the entire three-dimensional Cartesian space into four octants OI to OIV (not shown in the figure). This provides a three-dimensional geometric setting for the presentation of accounting events of all types. Since $q > 0$, the four octants in the bottom half of the three-dimensional space are not relevant.

Presenting Accounting Events: A Simple Case

In the three-dimensional (debit-credit-size) Cartesian geometric space, each element of an accounting event is shown as a point or line on a distinct coordinate. Together, the three points (lines) represent the accounting event and define a cubic space, which also represents the event. This space is a cube (or the cubic sphere therein) since the three coordinates are mutually perpendicular and $d = c = q$ under the double entry principles. Hereafter, this cube is called the double entry (DE) cube, and the sphere therein is called the DE sphere, of accounting events.[2] They are equivalent to the DE square and DE circle respectively of accounting events in Chap. 3. Conceptually, they reflect the duality and balance features of accounting events and the debit-credit symmetric nature of double entries or, more generally, the duality and balance features of double entry accounting.

Figure 4.1 gives an example of the DE cube and DE sphere respectively of an accounting event. This event is of type 1, the simplest type possible, and has a single account on both debit and credit sides. Thus, it has no component transactions. The event is assumed to increase the debit and credit accounts alike. Thus, the DE cube (DE sphere) lies in the first octant (OI) of the three-dimensional space. The height of the DE cube is equal to the monetary size of the event. The DE cube has eight vertices, one of which lies at the point of origin (zero monetary quantity).

In general, the location of the DE cube and DE sphere of type 1 accounting events in the three-dimensional geometric space depends on whether the events increase or decrease the debit and credit accounts. If an accounting event increases the debit and

[1] When a three-dimensional coordinate system (e.g., xyz) is set up, it is a convention in geometry that the second (y) coordinate axis is shown horizontally in front with its positive (negative) section on the right (left) side of the point of origin, while the first (x) coordinate axis is shown perpendicularly on the same plane with its positive section in front. This convention, however, is not followed (for convenience) in this chapter and all subsequent chapters. Instead, the debit and credit coordinates and their positive (+) and negative (−) sections as shown in Chap. 3 before are preserved.

[2] As with the DE square (DE circle) of accounting events in Chap. 3, the DE cube (DE sphere) of accounting events has different meanings, depending on how it is defined or used. It is a geometric presentation of accounting events in this chapter, the power of accounting events in Chap. 5, the DE vector space of accounting events in Chap. 6, and a geometric space represented by the DE matrix of accounting events in Chap. 7.

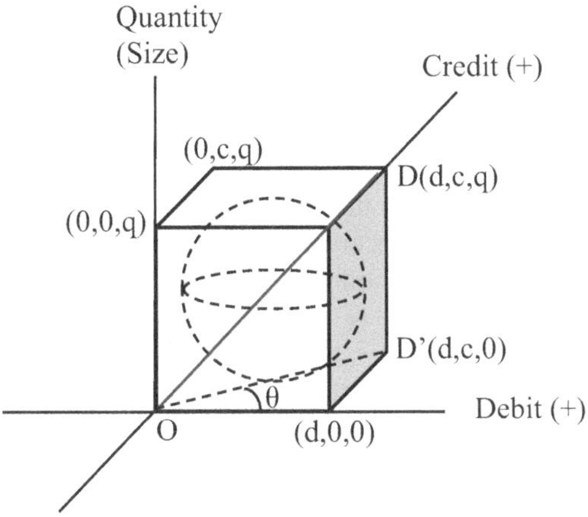

Fig. 4.1 A three-dimensional setting for the presentation of simple-type accounting events

credit accounts alike, the DE cube and DE sphere lie in OI. If an accounting event decreases the debit and credit accounts alike, the DE cube and DE sphere lie in OIII. If neither is the case, the DE cube and DE sphere lie in OII or OIV.

The size of the accounting event in monetary terms is equal to q. If measured geometrically in the three-dimensional space, it is equal to the size of the DE cube or the DE sphere therein. Each is computed as

$$
\begin{aligned}
Q_S &= d.c.q = d^3, c^3, \text{ or } q^3 \quad (d = c = q) \\
&= \left(\left|OD'\right| \text{ cosine } \theta\right).c.q \text{ or } d.\left(\left|OD'\right| \text{ sine } \theta\right).q \quad (\theta = 45^\circ)
\end{aligned}
$$

or

$$Q_S = 4\pi r^3/3 = 4\pi(q/2)^3/3 = \pi q^3/6$$

where |OD'| is the length of the line OD', θ is the angle of the line OD' with the debit (or credit) coordinate (θ = 45°), and r is the radius of the enclosed DE sphere. And d = |OD'| cosine θ under the law of cosine, and c = |OD'| sine θ under the law of sine. The slope of the line OD' is 1 as tangent 45° = 1. The diagonal OD of the DE cube has an angle of 35.2645° with the bottom face of the DE cube. Thus, it has a slope of 0.7071 as tangent 35.2645° = 0.7071.

Although different in magnitude, the two space-based measures above of the size of the accounting event are equivalent since they both are a geometric measure of the size of the same event. Thus, they can be converted to the same monetary size q of the accounting event.

$$q = {Q_S}^{1/3} \text{ or } (6Q_S/\pi)^{1/3}$$

And the ratio of the DE sphere to the DE cube is.

$$\left(\pi q^3/6\right)/Q_S = \left(\pi q^3/6\right)/d.c.q = \pi/6$$
$$= 0.5236$$

This ratio is comparable to the DE circle-to-DE square ratio (π/4) in Sect. 3.1, Chap. 3. As with the DE circle-to-DE square ratio in the two-dimensional (debit-credit) case there, the DE sphere-to-DE cube ratio is not affected by the complexity or simplicity of accounting events. Nor is it affected by the size of accounting events. It is the same, regardless of whether the accounting events are of type 1 or type 4 or whether their size is large or small.

Presenting Accounting Events: A Complex Case

The DE cube (sphere) in Fig. 4.1 represents accounting events of type 1, the simplest type possible. Many accounting events are not as simple. They are of type 2, 3, or 4 and have multiple accounts (entries) on the debit or credit side (or both sides). Such an accounting event is considered a bundle of component transactions, which are executed simultaneously as the event is executed. Thus, the debit, credit, and size elements respectively of the accounting event are not one continuous quantity but a multitude of discrete quantities. Nor is the DE cube or DE sphere of the event one continuous cubic space. It is made up of multiple component cubes (spheres) and rectangular cuboids. Also, the discrete cubic spaces may or may not lie in the same octant of the three-dimensional space since the multiple accounts on the debit or credit side, or both sides, of the accounting event may or may not change in the same direction.

Figure 4.2 gives an example of the DE cube of a complex accounting event in the three-dimensional geometric setting. This event is assumed to be of type 4 and has three component transactions (not shown in the figure). The DE cube shown looks similar in shape to, but is different in location from, the DE cube of the type 1 accounting event in the previous figure. It lies across the four octants of the three-dimensional space since the event is assumed to have component transactions and the multiple debit accounts and credit accounts respectively do not change in the same direction. The DE cube is divided into multiple component DE cubes and

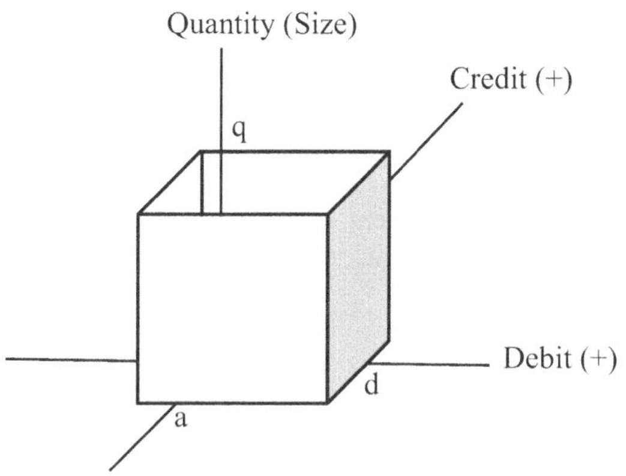

Fig. 4.2 A three-dimensional setting for the presentation of complex-type accounting events

rectangular cuboids although these are not shown in the figure (to keep the figure simple).

Despite the complexity of type 2, 3, or 4 accounting events, geometrically, their size can still be measured in the same way the size of type 1 accounting events was before, except that it is measured by the size of the component cubic spaces combined. This is shown below.

$$Q_S = d.c.q = \left(\sum |d_k|\right)\left(\sum |c_k|\right)\left(\sum q_k\right)$$

where d_k, c_k, and q_k are the debit, credit, and (monetary) size respectively of the kth component transaction of the accounting event in question. And $\sum|d_k| = \sum|c_k| = \sum q_k$ under the double entry principle. The components d_k and c_k respectively can be a decrease in monetary quantity, shown as a negative number in the three-dimensional space. Given this, and for notational convenience, all components of d and c are shown in absolute terms. This poses no problem since the size of a geometric space is an absolute scalar quantity. The size measure Q_S is not affected by the sign (+ or −) of d_k and c_k. The positive and negative signs of the component quantities only affect the location, not size, of the component cubic spaces. Also, when a component quantity is shown by a negative number, this does not mean that the monetary quantity shown by the number is intrinsically negative, as meant by the nonnegativity-of-money convention in accounting (see Sect. 2.2, Chap. 2). Although not shown to keep the figure simple, there is the DE sphere of the accounting event inside the DE cube, similar to the DE sphere in the previous figure. This sphere also represents the same accounting event, and its size is another space-based measure of the size of the event.

For accounting events of type 2, 3, or 4, the space-based measure Q_S of their size is not one continuous quantity but a multitude of discrete quantities. Measuring the size of such accounting events as such a multitude is consistent with the concept of quantity in general, which can be either one continuous magnitude or a multitude of discrete magnitudes (see Sect. 3.1, Chap. 3). It is also consistent with the notion of linear aggregation and the "whole-parts" relation—that the "whole" is the same as the sum of its "parts."

Since the DE sphere shown is equivalent to the DE cube, the size of type 2, 3, or 4 accounting events can also be measured by the size of their DE sphere. As in the case of type 1 accounting events before, the DE sphere-to-DE cube ratio for such accounting events is π/6, as shown below.

$$(\pi q^3/6)/Q_S = (\pi q^3/6)/d.c.q = \pi/6 = 0.5236$$

This ratio is equivalent to the DE circle-to-DE square ratio (π/4) in Sect. 3.1, Chap. 3. It is not affected by the complexity or simplicity of accounting events as indicated before. Thus, it is the same for all types of accounting events, regardless of whether the events are of type 1 or type 4.

4.2 Accounting Events in Three-Dimensional Space: Examples

How accounting events are presented according to the double entry principle in the three-dimensional (debit-credit-size) Cartesian space can be illustrated with various types of accounting events. Using the four transaction examples in Chap. 3, this section explains how it can be done.

Type 1 Events: A Single Account (Entry) on Both Sides

Consider the type 1 accounting event (in Chap. 3)—that a firm sells goods for \$100 in cash. This event has a single account (entry) on both debit and credit sides and increases the debit and credit accounts alike. Thus, it has no component transactions. Figure 4.3 presents the accounting event in the three-dimensional Cartesian space. The cash \$100 as the debit entry is a point on the positive (+) side of the debit coordinate and the revenue \$100 as the credit entry is a point (not seen) on the positive (+) side of the credit coordinate. And the monetary size \$100 of the accounting event is a point on the size coordinate. Together, the three points define a cube—the shaded DE cube and DE sphere of the accounting event. Since the event increases the debit and credit accounts alike, the DE cube and DE sphere lie in the first octant (OI) of the three-dimensional space. The bottom face of the cube is the DE square of the event.

The size of the accounting event in monetary terms is \$100, which is equal to the height of the DE cube (q = 100). If measured geometrically in the three-dimensional space, it is equal to the size of the DE cube or the DE sphere therein. Each is computed below.

$$Q_S = d.c.q = 100^3 vu^3$$

or

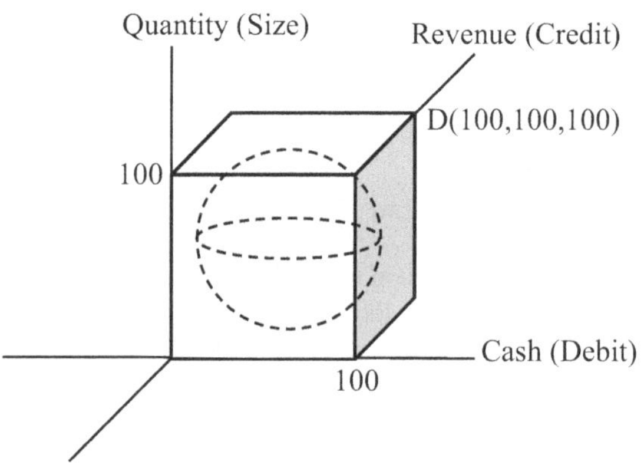

Fig. 4.3 Presenting a type 1 accounting event in the three-dimensional (debit-credit-size) space

$$Q_S = 4\pi r^3/3 \text{ or } 4\pi 50^3/3 = 523,598.78 \text{ vu}^3$$

where vu^3 denotes a "volume unit" (e.g., cubic inch). Although different in magnitude, the two space-based measures of the size of the accounting event are equivalent since they both are a geometric measure of the size of the same accounting event. And the DE sphere-to-DE cube ratio is

$$(4\pi r^3/3)/Q_S = 4\pi(q/2)^3/3q^3 = \pi/6 = 0.5236$$

As the two geometric measures of the size of the accounting event are equivalent, they can be converted to the monetary size of the accounting event:

$$q = (Q_S)^{1/3} = \left(100^3\right)^{1/3} = 100$$

or

$$q = 2r = (2)[(3)(523,598.78)/4\pi]^{1/3} = 100$$

This accounting event is of type 1 and increases the debit and credit accounts. Thus, the DE cube and DE sphere of the event lie in OI. If an accounting event is of the same type and decreases the debit and credit accounts alike, the DE cube and DE sphere of the event lie in OIII. If neither is the case, they lie in OII or OIV. Yet the geometric measure Q_S of the size of the accounting event is still computed in the same way.

Type 2 Events: Multiple Accounts (Entries) on the Debit Side

Consider the type 2 accounting event (in Chap. 3)—that a firm sells goods for $100 by receiving $60 in cash and holding $40 on account for future collection. This event has two debit accounts (cash and account receivable), both as an increase, and one credit account, also as an increase. Thus, it has two component transactions, one for the cash part of it and the other for the account-receivable part. Figure 4.4 shows the accounting event in the three-dimensional Cartesian space. The large unshaded cube is the DE cube of the accounting event (the related DE sphere is not shown to keep the figure simple). Since the accounting event increases the debit and credit accounts alike, the DE cube lies in OI of the three-dimensional space. As the event has two component transactions, the DE cube is divided into two small DE cubes and six rectangular cuboids. These cubes and cuboids are determined by slicing the DE cube of the accounting event horizontally and vertically along the dashed line on each face of it. The small shaded cube in the lower-left corner of the large DE cube represents the cash part ($60) of the accounting event, and the small shaded cube in the upper-right corner represents the account-receivable part ($40). The two small cubes are connected at a common vertex inside the large DE cube.

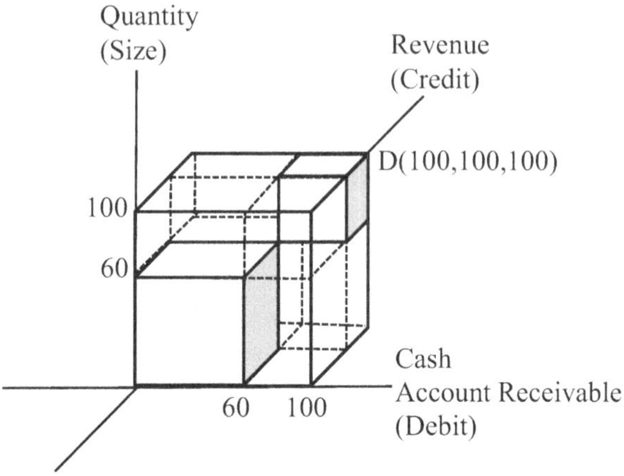

Fig. 4.4 Presenting a type 2 accounting event in the three-dimensional (debit-credit-size) space

The size of the accounting event, if measured in monetary terms, is \$100, which is equal to the height of the DE cube of the event ($q = 100$). If measured geometrically, it is equal to the size of the large unshaded DE cube or the DE sphere therein (not shown). Each is computed as

$$Q_S = d.c.q = (60 + 40)(60 + 40)(60 + 40) = 100^3 vu^3$$

or

$$Q_S = 4\pi r^3/3 \text{ or } 4\pi 50^3/3 = 523,598.78 \text{ vu}^3$$

The space-based measure Qs above of the size of the accounting event equals the size of the two small shaded cubes combined plus the size of the six unshaded rectangular cuboids combined. Algebraically, this can be seen by expanding the first expression of Q_S above. The size of each small DE cube is equal to the direct product of the debit and credit of the related component transaction (60.60 and 40.40), and the size of each rectangular cuboid is equal to the cross product of the debit of one component transaction and the credit of the other (e.g., 60.40 or 40.60). As the accounting event has component transactions, its size is not one continuous quantity but a multitude of discrete quantities. The size of the accounting event can also be measured by the size of the DE sphere of the event as shown.

Since the DE cube and DE sphere of the accounting event are equivalent, the two size measures are compatible. The DE sphere-to-DE cube ratio is

$$(4\pi r^3/3)/Q_S = 4\pi(q/2)^3/3q^3 = \pi/6 = 0.5236$$

which is equal to the same ratio for the previous (type 1) accounting event. The ratio is the same since it is not affected by the types of accounting events as indicated before.

Since the two space-based measures of the size of the accounting event are equivalent, they can be converted to the same monetary size of the event:

$$q = (Q_S)^{1/3} = \left(100^3\right)^{1/3} = 100$$

or

$$q = 2r = (2)[(3)(523,598.78)/4\pi]^{1/3} = 100$$

The type 2 accounting event above has two debit accounts and one credit account and increases each of them. If an accounting event is of the same type but decreases one of the two debit accounts, the DE cube lies across OI and OII of the three-dimensional space. And one of the two component DE cubes lies in OI, and the other in OII. If it decreases both of the debit accounts, the DE cube lies in OII. Regardless, the accounting event in each case can be shown as a DE cube (DE sphere) and its size is measured by the size of the DE cube (DE sphere), computed in the same way.

Type 3 Events: Multiple Accounts (Entries) on the Credit Side

To see how this type of accounting events can be examined in the three-dimensional space, refer back to the type 3 transaction example (in Chap. 3)—that a firm purchases goods for \$100 by paying \$60 in cash and holding \$40 on account. Like the type 2 accounting event before, this accounting event has two component transactions, one for the cash part and the other for the account-payable part. Figure 4.5 presents the accounting event in the three-dimensional Cartesian space. The large unshaded cube is the DE cube of the accounting event. The inventory asset \$100 as a debit account is shown on the positive (+) side of the debit coordinate. The cash \$60 as a credit account is shown as −60 on the negative side of the credit

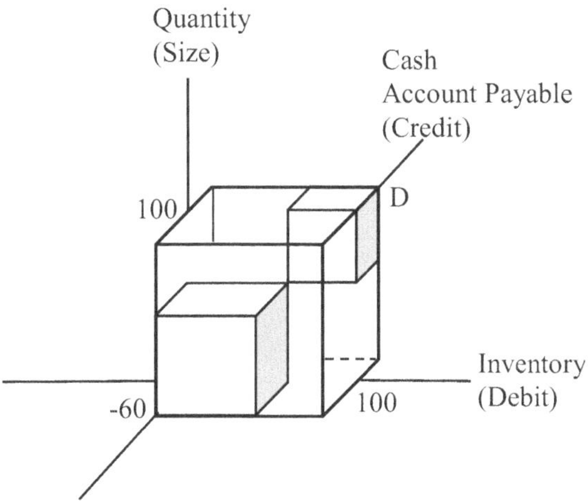

Fig. 4.5 Presenting a type 3 accounting event in the three-dimensional (debit-credit-size) space

coordinate as it is credited to the cash account as a decrease in the journal entry. Because of this, the unshaded DE cube of the accounting event lies across OI and OIV. As the event has two component transactions, the DE cube is divided into two small cubes and six rectangular cuboids (although not shown to keep the figure simple). The shaded small DE cube in OIV represents the cash part of the accounting event, and the one in OI represents the account-payable part. The two small DE cubes are connected at a common vertex inside the large DE cube. And some of the six cuboids lie in QI and others in QIV.

If measured in monetary terms, the size of the accounting event equals \$100, which is equal to the height of the large unshaded DE cube (q = 100). If measured geometrically in the three-dimensional space, it is equal to the size (volume) of the DE cube or DE sphere. Each is computed as

$$Q_S = d.c.q = (60 + 40)(40 + |-60|)(60 + 40) = 100^3 vu^3$$

or

$$Q_S = 4\pi r^3/3 \text{ or } 4\pi 50^3/3 = 523,598.78 \text{ vu}^3$$

Since the decrease in cash of \$60 is shown as a negative number (−60), it is shown in absolute value in the expression above. This does not affect Q_S as a measure of the size of the DE cube since a geometric space is an absolute scalar quantity. The size of the accounting event, measured by Q_S, equals the size of the two small DE cubes combined plus the size of the six rectangular cuboids combined.[3] The size of each small cube is equal to the direct product of the debit and credit of the related component transaction, and the size of each rectangular cuboid is equal to the cross product of the debit (credit) of one component transaction and the credit (debit) of the other. As in the case of the type 1 and 2 accounting events before, the DE sphere-to-DE cube ratio is $4\pi(q/2)^3 / 3q^3 = \pi/6 = 0.5236$, which is the same as the ratio for the previous transactions.

The two measures Q_S above of the size of the accounting event are equivalent as they both concern the same accounting event. Thus, each can be converted to the same monetary size (\$100) of the event as follows:

$$q = (Q_S)^{1/3} = \left(100^3\right)^{1/3} = 100$$

or

[3] To see this algebraically, the expression of Q_S can be expanded as: 60.40.60 + 60.40.40 + 60.|−60|.60 + 60.|−60|.40 + 40.40.60 + 40.40.40 + 40.|−60|.60 + 40.|−60|.40 = 100^3 vu^3. The third term (60.|−60|.60) equals the size of the shaded DE cube in OIV for the cash part of the accounting event, and the sixth term (40.40.40) equals the size of the shaded DE cube in OI for the account-payable part. Each of the six cross products equals the size of the related rectangular cuboid.

$$q = 2r = (2)[(3)(523,598.78)/4\pi]^{1/3} = 100$$

The type 3 accounting event above increases one of the credit accounts, the account payable, and decreases the other, the cash account. If an accounting event is of type 3 and increases the debit and credit accounts alike, the DE cube and related component DE cubes lie in OI of the three-dimensional space. If an accounting event is of the same type and decreases the debit and credit accounts alike, the DE cube and related component DE cubes lie in OIII.

Type 4 Events: Multiple Accounts (Entries) on Both Sides

More complex than the previous three types are accounting events of type 4 as they have multiple accounts (entries) on both debit and credit sides and the multiple accounts on each side may not change in the same direction. To see how such accounting events can be presented in the three-dimensional space, consider the type 4 accounting event (in Chap. 3)—that a firm recognizes an interest expense of $330 on a bond liability and pays the cash of $350 for the interest on the bond liability. The interest expense recognized includes the current-period bond discount amortization of $100. The cash paid includes a previously accrued but unpaid interest of $120 and a newly accrued interest of $230 for the current period. Given the above, the accounting event has three component transactions (see Sect. 3.2, Chap. 3).

Figure 4.6 presents the accounting event in the three-dimensional Cartesian space. The shaded cube is the DE cube of the event. The interest expense $330 is a point on the positive (+) side of the debit coordinate. The payment of the interest payable $120 is a point (but not seen) on the negative (−) side of the debit coordinate since it is debited to the interest-payable account as a decrease. The cash $350 paid is

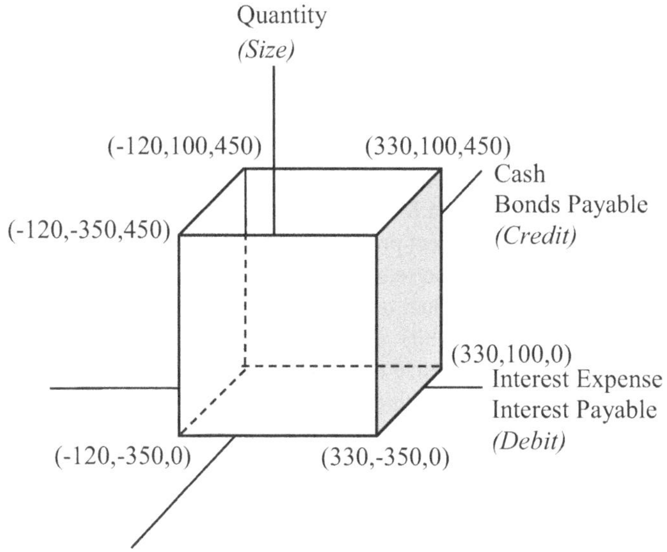

Fig. 4.6 Presenting a type 4 accounting event in the three-dimensional (debit-credit-size) space

a point on the negative (−) side of the credit coordinate as it is credited to the cash account as a decrease. It includes the newly accrued current-period interest \$230 and the interest payable \$120. The bond discount amortization \$100 increases the bond liability so that it is a point on the positive (+) side of the credit coordinate. Given all this, the shaded DE cube of the accounting event lies across four octants of the three-dimensional space. The top face of the DE cube has four vertices: (330,100,450), (−120,100,450), (−120,−350,450), and (330,−350,450). The corresponding vertices of the bottom face of the DE cube are (330,100,0), (−120,100,0), (−120,−350,0), and (330,−350,0).

As the accounting event has three component transactions, the shaded DE cube is divided into three small DE cubes—one for each component transaction—and 24 rectangular cuboids (not shown to keep the figure simple). If shown, the DE cube for the increase in the bond liability due to the amortization of the bond discount (\$100) would lie in OI, the DE cube for the payment of the unpaid interest (\$120) in OIII, and the DE cube for the payment of the current-period interest (\$230) in OIV. There is no component DE cube in OII since the accounting event has no component transaction that has a debit account as a decrease and a credit account as an increase. The DE cube of the accounting event lies across the four octants of the three-dimensional space since the multiple accounts on each side (debit and credit) do not change in the same direction, increase or decrease.

The size of the accounting event, if measured in monetary terms, is \$450, equal to the height of the DE cube of the event (q = 450). If measured geometrically, it is equal to the size (volume) of the DE cube or the DE sphere therein. Each is computed as

$$\begin{aligned} Q_S &= (100 + 230 + |-120|)(100 + |-120| + |-230|)(450) \\ &= 450^3 \text{vu}^3 \end{aligned}$$

or

$$Q_S = 4\pi r^3/3 = 4\pi 225^3/3 = 47,712,898.13 \text{ vu}^3$$

The size of the DE cube, measured by Q_S, equals the size of the three small DE cubes combined plus the size of the 24 rectangular cuboids combined.[4] The size of each small DE cube is equal to the direct product of the debit, credit, and size elements of the related component transaction (e.g., 100.100.100). The size of each rectangular cuboid is equal to the cross product of the debit (credit) element of one component transaction and the credit (debit) element of another component transaction, multiplied by the corresponding size element (e.g., 230.100.100).

As the accounting event has three component transactions, the space-based measure Q_S of its size is not one continuous quantity but a multitude of discrete

[4]To see this algebraically, the expression of Q_S can be expanded as done for the type 3 accounting event in the previous footnote.

quantities. Yet the DE sphere-to-DE cube ratio is equal to the same ratio for the previous accounting events: $(4\pi r^3/3)/Q_S = 4\pi(q/2)^3/3q^3 = \pi/6 = 47{,}712{,}938.43/450^3 = 0.5236$. It is the same even if this event is much larger in size and more complex in nature. Again, this is because the DE sphere-to-DE cube ratio is not affected by the complexity of accounting events, nor by their monetary size, as indicated before.

Since the two space-based measures above of the size of the accounting event are equivalent, each can be converted to the same monetary size ($450) of the event:

$$q = (Q_S)^{1/3} = \left(450^3\right)^{1/3} = 450$$

or

$$\begin{aligned} q &= 2r = (2)(3Q_S/4\pi)^{1/3} = (2)[(3)(47,712,898.13)/4\pi]^{1/3} \\ &= 450 \end{aligned}$$

4.3 Features of the Double Entry Cube

The DE cubes (and DE sphere) of the accounting events discussed thus far in this chapter have interesting features, which are comparable to the features of the DE squares (and DE circle) of those accounting events in Chap. 3. As with the DE squares there, the DE cubes reflect the duality and balance features of the accounting events and the contrasting yet symmetric nature of double entries in the three-dimensional (debit-credit-size) Cartesian space. As such, they are not just any cube but a special kind of cube. Reflecting the notion of duality and balance, they have six equal square faces. And their sides and heights lie on, or are in parallel with, the corresponding coordinates of the three-dimensional space. That is, they do not tilt toward or away from any particular coordinate. Nor do they lie away from the point of origin (zero monetary quantity) of the three-dimensional space. Rather, they have a vertex at the point of origin or have the point of origin as an inner point. Or they have the point of origin on the debit or credit side of their bottom face of the DE cube. Also, they lie in any of the four octants of the three-dimensional space, across the two adjacent octants—(OI, OII), (OII, OIII), (OIII, OIV), and (OI, OIV), or across all four octants (OI to OIV). However, they cannot lie across any three octants, while still having those features of the DE cubes described above.

Given the features of the DE cubes, not all cubes (or round spheres therein) or similar geometric shapes in the three-dimensional space can be a DE cube (DE sphere). Figure 4.7 gives an example of one such cube. This cube looks like a DE cube but is not since it does not have the features of a DE cube described above. It has six equal square faces and lies across the four octants of the three-dimensional space. It has the point of origin as an inner point of its bottom face. The sides of the cube are the same and are also equal to its height. However, none of the sides of the cube is along, or in parallel with, the debit or credit coordinate. As such, the posture of the cube is not consistent with the features of the DE cube of accounting events.

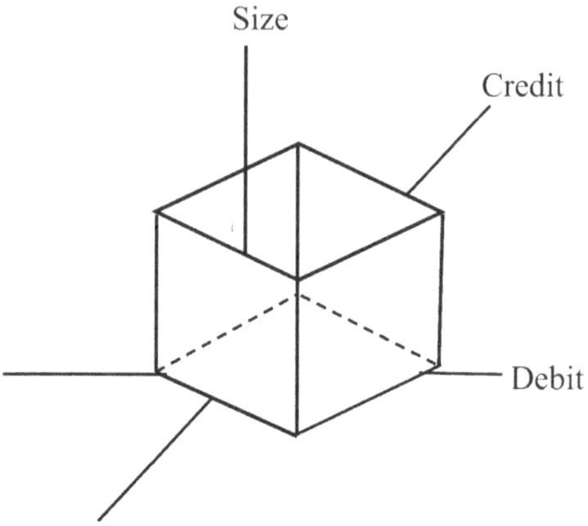

Fig. 4.7 A cube that is *not* a double entry (DE) cube in the three-dimensional (debit-credit-size) space

Nor is it consistent with the duality and balance features of accounting events or the debit-credit contrasting yet symmetric nature of double entries. There is no accounting event that can geometrically fit into the cube.

4.4 Accounting Events as Lines and Points

It is possible to present accounting events as a line and a point respectively, as well as a space, in the three-dimensional (debit-credit-size) Cartesian space as it was in the two-dimensional (debit-credit) space in Sect. 3.4, Chap. 3, before. This section discusses the line- and point-based presentation of accounting events and the line- and point-based measurement of their size in the three-dimensional space. It does so with the two transaction examples, type 1 and type 4, seen in the previous section.

The line representing an accounting event is the diagonal of the DE cube of the event, and the point is the diagonal vertex of the DE cube. The former is the double entry (DE) line of the accounting event, and the latter is the double entry (DE) point, in the three-dimensional space. They are equivalent to the DE line and DE point respectively in Chap. 3. Each of them is an ordered array of triple numbers—d, c, and q. For a given accounting event, the DE line and DE point are equivalent. They are also equivalent to the DE cube of the event since they all represent the same accounting even and reflect the same duality and balance features of the event. As such, if one of them is known or determined, the others are also known or determined.

Example 1 Refer back to the type 1 accounting event in the previous section. Figure 4.8, adapted from Fig. 4.3, presents the event in the three-dimensional Cartesian space. The shaded cube is the DE cube of the event. The point d on the debit coordinate represents the cash of $100 as an increase, and the point c (not seen) on the credit coordinate represents the revenue of $100 as an increase. The point q on the size coordinate represents the monetary size $100 of the event. The diagonal (line OD) of the DE cube is the DE line of the accounting event, and the diagonal vertex

Fig. 4.8 The line- and point-based presentation of a type 1 accounting event in the three-dimensional (debit-credit-size) space

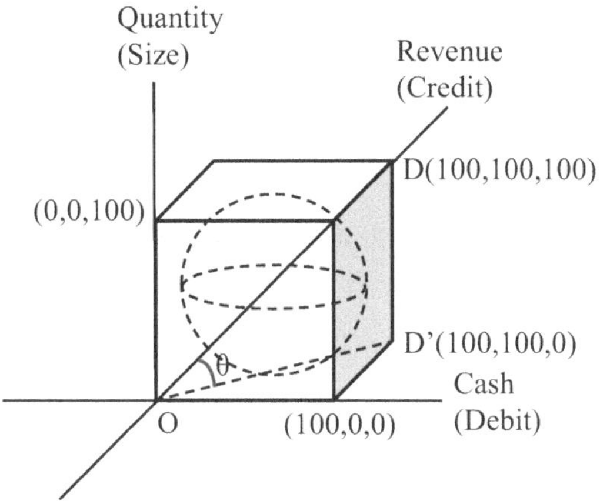

(point D) is the DE point. Each of these reflects the same duality and balance features of the accounting event as the DE cube does. The DE line has an angle of 35.2645° (θ = 35.2645°) with the bottom plane of the DE cube and defines the right triangle ODD'. And the slope of the DE line is 0.7071 as tangent 35.2645° = 0.7071.[5] The line OD' on the bottom face of the DE cube is also the DE line of the accounting event. It reflects the same duality and balance features of the event. It has an angle of 45° with the debit and credit coordinates.

If measured geometrically in the three-dimensional space, the size of the accounting event is equal to the length of the DE line or the distance of the DE point from the origin (point O). Given the right triangle ODD', it can be computed according to the Pythagorean theorem, or the law of cosine and the law of sine respectively, as

$$\begin{aligned} Q_L &= \left(|DD'|^2 + |OD'|^2\right)^{1/2} = \left(|DD'|^2 + |Od|^2 + |dD'|^2\right)^{1/2} \\ &= \left(q^2 + d^2 + c^2\right)^{1/2} = \left(100^2 + 100^2 + 100^2\right)^{1/2} \\ &= 173.2051 \end{aligned}$$

or

$$\begin{aligned} Q_L &= |OD'|/\text{cosine } \theta = \left(|Od|^2 + |dD'|^2\right)^{1/2}/\text{cosine } \theta \\ &= \left(100^2 + 100^2\right)^{1/2}/\text{cosine } \theta = 100\sqrt{2}/0.81649547 \\ &= |DD'|/\text{sine } \theta = |DD'|/\text{sine } \theta = 100/0.57735 \\ &= 173.2046 \end{aligned}$$

[5]The slope of the DE line is defined as |DD'| / |OD'| = |DD'| / $(|Od|^2 + |dD'|^2)^{1/2}$ = 100 / $(100^2 + 100^2)^{1/2}$ = 0.7071, which is roughly equal to tangent 35.2645°.

where $\theta \approx 35.2645°$. The line-based measure Q_L of the size of the accounting event is equal to the distance of the DE point D from the point of origin. It is equivalent to the space-based measure Q_S (DE cube) of the size of the event in Sect. 4.2. The relation between Q_S and Q_L is shown below. Based on this, both Q_S and Q_L can be converted to the same monetary size ($100) of the accounting event:

$$\begin{aligned} Q_S &= \text{d.c.q} = (Q_L \text{ cosine } \theta)(Q_L \text{ cosine } \theta)(q) \\ &= (173.2051 \text{ cosine } \theta)^2 (100) = 100^3 \text{ vu}^3 \end{aligned}$$

$$q = \left(100^3\right)^{1/3} = 100$$

where $\theta \approx 54.7355°$, the angle of the DE line OD of the accounting event with the debit and credit coordinates. And

$$\begin{aligned} q^2 &= |DD'|^2 = {Q_L}^2 - |OD'|^2 = {Q_L}^2 - \left(|Od|^2 + |dD'|^2\right) \\ q &= \left[173.2051^2 - \left(100^2 + 100^2\right)\right]^{1/2} = 100 \end{aligned}$$

or

$$\begin{aligned} Q_L &= q/\text{sine } \theta \\ q &= Q_L \text{sine } \theta \\ &= |DD'| = 173.2051 \text{ sine } \theta \\ &= (173.2051)(0.5773518) = 100 \end{aligned}$$

where $\theta \approx 35.2645°$.

In general, the line- and point-based measures of the size of type 1 accounting events in the three-dimensional space is computed as.

$$Q_L = \left(d^2 + c^2 + q^2\right)^{1/2} = d\sqrt{3}, c\sqrt{3}, \text{ or } q\sqrt{3} \quad (d = c = q)$$

or

$$Q_L = q/\text{sine } \theta$$

or

$$\begin{aligned} Q_L &= \left(d^2 + c^2\right)^{1/2}/\text{cosine } \theta \\ &= d\sqrt{2}/\text{cosine } \theta \text{ or } c\sqrt{2}/\text{cosine } \theta \end{aligned}$$

where $\theta \approx 35.2645°$.

Example 2 Presenting type 4 accounting events and measuring their size based on their DE line and DE point respectively in the three-dimensional space is somewhat complex. This is because such accounting events have multiple accounts on both

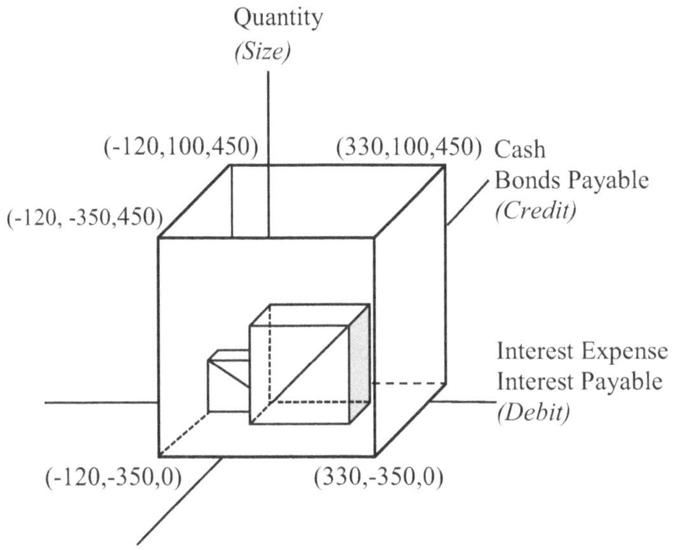

Fig. 4.9 The line- and point-based presentation of a type 4 accounting event in the three-dimensional (debit-credit-size) space

debit and credit sides and the multiple accounts on each side may not change in the same direction, increase or decrease. To see how such accounting events can be presented as a line and a point respectively in the three-dimensional space, refer back to the type 4 accounting event—that a firm recognizes an interest expense of $330, pays an interest of $350, and amortizes a discount of $100, all relating to a bond liability. This event has three component transactions. Figure 4.9, adapted from Fig. 4.6, graphically presents the event in the three-dimensional space. The large (unshaded) cube is the DE cube of the accounting event. Since the event does not change the debits and credits of these component transactions in the same direction, the three small DE cubes of the component transactions lie in the three different octants.[6] And the DE cube of the event lies across the four octants of the three-dimensional space. Each small DE cube represents a component transaction. And its diagonal is the DE line, and its diagonal vertex is the DE point, of the component transaction. Given the three component transactions, the DE line of the accounting event is divided into three different DE lines, one for each component transaction, and the DE point is represented jointly by the three DE points of the component transactions.

If measured based on a line or point, the size of the accounting event equals the length of the DE lines (diagonals) of the three component transactions combined or

[6]The small DE cube for the increase in the bond liability due to the bond discount amortization $100 lies in OI, behind the large shaded DE cube.

the distance of their DE points (diagonal vertices) combined. Each of these size measures is computed below.

$$
\begin{aligned}
Q_L &= \left[q^2 + \left(|d|^2 + |c|^2\right)\right]^{1/2} \\
&= \left[450^2 + \left(450^2 + 450\right)\right]^{1/2} \\
&= 450\sqrt{3} = 779.4229
\end{aligned}
$$

or

$$
\begin{aligned}
Q_L &= 450\sqrt{2}/\text{cosine } \theta \text{ or } 450/\text{sine } \theta \\
&= 779.4229
\end{aligned}
$$

or

$$
\begin{aligned}
Q_L &= \left({d_1}^2 + {c_1}^2 + {q_1}^2\right)^{1/2} + \left({d_3}^2 + {c_3}^2 + {q_3}^2\right)^{1/2} + \left({d_4}^2 + {c_4}^2 + {q_4}^2\right)^{1/2} \\
&= \left(100^2 + 100^2 + 100^2\right)^{1/2} + \left(|-120|^2 + |-120|^2 + 120^2\right)^{1/2} \\
&\quad + \left(230^2 + |-230|^2 + 230^2\right)^{1/2} \\
&= 779.4229
\end{aligned}
$$

or

$$
\begin{aligned}
Q_L &= 100\sqrt{2}/\text{cosine } \theta + |-120|\sqrt{2}/\text{cosine } \theta + 230\sqrt{2}/\text{cosine } \theta \\
&= 779.4229
\end{aligned}
$$

where $\theta \approx 35.2645°$ and the subscripts 1, 3, and 4 denote the octant I, III, and IV respectively of the three-dimensional space. Q_L in absolute terms is equal to the combined length of the three DE lines or the combined distance of the three DE points from the point of origin. Measuring the size of the accounting event as shown is reasonable since the length of a line and the distance of a point are an absolute scalar quantity.

The alternative line- and point-based measures above of the size of the accounting event equal the length of the diagonal of the large unshaded DE cube of the event. They are equivalent and can be converted to the same monetary size of the event:

$$
\begin{aligned}
q^2 &= {Q_L}^2 - \left(|d|^2 + |c|^2\right) = 779.4229^2 - \left(450^2 + 450^2\right) \\
&= 202,500 \\
q &= 450
\end{aligned}
$$

In general, for type 2, 3, or 4 accounting events having component transactions, the line-based measure of the size of the events equals the diagonal of the component

DE cubes combined. And the point-based measure of their size equals the distance of the component DE points combined. Each is shown in general notations as

$$\begin{aligned} Q_L &= \left[\left(\sum |d_k|\right)^2 + \left(\sum |c_k|\right)^2 + \left(\sum q_k\right)^2\right]^{1/2} \\ &= \sqrt{3}\sum |d_k|, \sqrt{3}\sum |c_k|, \text{or } \sqrt{3}\sum q_k \end{aligned}$$

or

$$\begin{aligned} Q_L &= q_1/\text{sine } \theta + q_2/\text{sine } \theta + \ldots + q_k/\text{sine } \theta \\ &= \sum q_k/\text{sine } \theta \end{aligned}$$

where d_k, c_k, and q_k are the debit, credit, and size elements of the kth component transaction of the type 2, 3, or 4 accounting event in question and $\theta \approx 35.2645°$. And $\sum|d_k| = \sum|d_k| = \sum|d_k|$ under the double entry principle. Some component quantities of d and c respectively may be a decrease, thus shown as a negative number in the three-dimensional space. Given this possibility, and for notational convenience, all component quantities are shown in absolute terms. When computed as shown, Q_L is not one continuous magnitude but a multitude of discrete magnitudes. Measuring the size of the accounting event as such a multitude is consistent with the notion of linear aggregation and the "whole-parts" relation indicated earlier.

For a given accounting event, the line-based measure Q_L above of its size is equivalent to the corresponding line-based measure of its size in the two-dimensional case in Sect. 3.4, Chap. 3. Hence, they can be converted to the same monetary size of the event. If, for example, an accounting event has two component transactions, the monetary size of the event is equal to.

$$\begin{aligned} q &= \left[\sqrt{3}\left(q_1{}^2 + q_2{}^2\right)\right]^{1/3} = \text{d or c} \\ &= \left[\sqrt{3}(|d_1| + |d_2|)\right]^{1/3} \text{ or } \left[\sqrt{3}(|c_1| + |c_2|)\right]^{1/3} \\ &= \sum_k |d_k| = \sum_k |c_k| = Q_L/\sqrt{3} \end{aligned}$$

where $d_1 = c_1 = q_1$ and $d_2 = c_2 = q_2$ under the double entry principle.[7]

[7]For example, the monetary size of the type 4 transaction example is computed as $q = Q_L / \sqrt{3} = 779.4229 / \sqrt{3} = 450$.

4.5 Relative Merits of the Three-Dimensional Approach

Before closing the chapter, some comments are in order regarding the relative merits of the three-dimensional geometric approach to double entry accounting. First, the three-dimensional (debit-credit-size) setting, relative to the one- and two-dimensional ones seen in Chaps. 2 and 3, fits conceptually better for the multidimensional nature of accounting events. Accounting events are multidimensional as they have three dimensions—debit, credit, and size, represented by their three elements d, c, and q. In the three-dimensional space, there exists a dimension for each of those dimensions, and each of those elements is shown as a point or line on a distinct dimension. This is not possible in the one- and two-dimensional settings as seen in Chaps. 2 and 3 before. Thus, the three-dimensional setting fits conceptually better for presenting accounting events geometrically and offers more (visual) insights into the duality and balance features of accounting events.

Second, the DE cube of accounting events can be seen as a square box in the three-dimensional space. It can be moved around from one location to another, without changing its size. The square box at each location represents an accounting event. Since there are an almost countless number of ways to move it around from one location to another in the three-dimensional space, there are an equally countless number of accounting events that can be presented in the entire three-dimensional space. If the size of the square box at each location is changed as well, the total number of accounting events that can be shown as a square box in the entire three-dimensional space increases exponentially and rises beyond a level of any easy calculation.

Third, alternative geometric presentations of accounting events with spaces, lines, and points respectively in the three-dimensional space make it possible to examine various properties of accounting events geometrically. Accounting events, unlike their debit and credit entries, have not only multiple dimensions (debit, credit, and size) but also a multitude of properties, some of which are nonmonetary and, hence, beyond monetary measurement and interpretation. They are dimensionality, duality, force, power, informativeness, and quality, among others. Given this, a single way of presenting accounting events, and a single way of measuring their size, namely in monetary terms only, is not enough to gain insights into such properties of accounting events. Multiple ways of presenting accounting events, and multiple ways of measuring their size, in geometric spaces help gain good insights into those properties.

Finally, presenting accounting events as lines, points, and spaces in the three-dimensional space makes it possible to examine accounting events and explore the features of double entry accounting in multidimensional vector spaces. A line, point, and space respectively are a vector or a set of vectors in such vector spaces. And the line, point, and space representing accounting events can also be written as a vector or a set of vectors, or a matrix, in such vector spaces. These are discussed in Chaps. 6 and 7 later.

The Special Properties of Accounting Events 5

Accounting events have not only three intrinsic elements (debit, credit, and size) but also a multitude of properties, some of which are nonmonetary and are hence beyond monetary measurement and interpretation. Examples of such properties include dimensionality, duality, force, power, quality, and informativeness, among others. The dimensionality and duality of accounting events were examined in Chaps. 1 and 2. This chapter discusses two additional properties—force and power—of accounting events in two- and three-dimensional geometric settings.[1] It does so with the transaction examples in Chap. 3 for illustrations. Research on force and power is well established in other fields of study such as physics, engineering, and sociology, to name a few. They are often employed to explain how the properties of an object, a thing, or a phenomenon exist or work. Such, however, is not the case in accounting although accounting events have force and power as their intrinsic properties.

5.1 The Forces of Accounting Events

The Concept of Force

As a basic concept, force means a property that causes an object, a thing, or a phenomenon to move or change. It has magnitude and direction. When force is defined in such general terms, it has different meanings, depending on how it is seen or works. If a force concerns a physical object, it is considered a physical property. The force of a car engine to move the car is an example. Force can also be a nonphysical property, which concerns something intangible. In sociology, for example, force is considered a property that affects the behavior or mind of people.[2] Although nonphysical, it can affect the behavior or mind of people and cause them to

[1] The other nonmonetary properties of accounting events are not examined in this book but left out as topics for future work (see Chap. 10).

[2] See https://www.sociologyguide.com/socio-notes/social-force.php (November 2020).

© The Author(s), under exclusive license to Springer Nature Switzerland AG 2025
B. T. Ro, *The Geometry of Accounting*, Springer Texts in Business and Economics,
https://doi.org/10.1007/978-3-031-88892-2_5

take an action, which can lead to a political or social movement or a political change, for example. The force of Nature, which causes natural things or phenomena to change, is another example of nonphysical force.

Regardless of how the concept of force is defined or seen, there is one thing in common; force has magnitude and direction as its intrinsic attributes. The magnitude of a force is a scalar quantity, and its direction means the direction in which the force acts or points. While a force has a direction, its magnitude is independent of the direction. Thus, the magnitude of a force is the same, regardless of whether the underlying force acts to the left or to the right, all else equal.

The Forces of Accounting Events and Their Types

Accounting events are a dynamic and flow concept in the sense that they affect the stock balance of the capital resources (capital) of a business entity (firm) and the composition of the sources and uses of those resources. As such, they have a force, regardless of whether they are arm's-length transactions or not. The sources and uses of capital resources and their stock relationship are shown in a mathematical equation called the fundamental accounting model: assets (A) = liabilities (L) + owner equity (Eq). Hence, it can be said that an accounting event has a force as it affects the model, namely the sources and uses of the capital resources of a firm, in one way or another. This force is nonphysical since accounting events as well as the fundamental accounting model are nonphysical.

The force of accounting events is not a force of single type since not all accounting events affect the fundamental accounting model in the same way. Figure 5.1 graphically shows three different types of the force of accounting events. Each double-headed arrow represents a particular type of the force of accounting events. The horizontal double-headed arrow represents the force of accounting events that affects both sides of the fundamental accounting model—the sources and uses of the capital resources of a business entity (a firm)—and does so to the full extent of their monetary size. If, for example, a firm sells goods for \$100 in cash or purchases goods for \$100 on account, this accounting event has a force that affects both sides of the model by \$100. Hereafter, this type of force is called the type B force of accounting events.

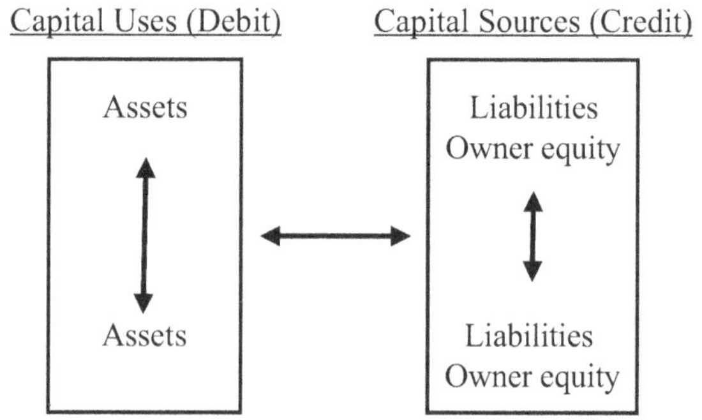

Fig. 5.1 Types of the force of accounting events and the fundamental accounting model

Unlike such accounting events, some accounting events affect only one side of the fundamental accounting model. Accounting events having this type of force affect only the sources or uses, not both, of the capital resources of a firm. This force is called the type O force of accounting events hereafter. It is shown by the vertical double-headed arrows in the figure. Since it affects only one side of the capital resources, it does not change the monetary size of the capital resources of the firm. If, for example, a firm collects an account receivable in cash, this accounting event has a type O force. It affects only the left side of the fundamental accounting model, namely the composition of the uses of capital resources. It does not affect the sources, nor the overall monetary size, of the capital resources.

There is still another type of force, which is a mixture of the two force types above. This force is referred to as the type M force of accounting events. Accounting events having this type of force affect both sides of the fundamental accounting model, hence the sources and uses of capital resources, but not to the full extent of their monetary size. If, for example, a firm purchases some goods for \$100 partly in cash and partly on account, this accounting event has a force of type M since it increases both the inventory asset and the account payable, hence the monetary size of the capital resources of the firm, but not by \$100.

Every accounting event, no matter how complex or simple it is, has one of the three types of force. Accounting events of type 1 have a type B or O force since they have no component transaction. By nature, they cannot have a type M force. Accounting events of type 2, 3, or 4 can have any of the three types of force depending on how they change the multiple debit and credit accounts—in the same direction or in different directions. Such accounting events have component transactions, and each component transaction has a force of its own, which is of type B or O. As a component transaction has a single account on both debit and credit sides, it is similar to a type 1 accounting event. Therefore, by nature, a component transaction cannot have a type M force. Having multiple component transactions, type 2, 3, or 4 accounting events have a multitude of discrete forces. These discrete forces may or may not act in the same direction. If they do (increase or decrease), the accounting events have a type B or O force. Otherwise, they have a type M force since they affect the fundamental accounting model but not to the full extent of their monetary size.

What the above means is that the types of the force of accounting events are inherently related to the types of accounting events (see Sect. 2.4, Chap. 2). This is shown in journal entry form [debit credit] below.

Type B force: $[+A \quad +L], [+A \quad +Eq], [-L \quad -A]$, or $[-Eq \quad -A]$

Type O force: $[+A \quad -A], [-L \quad +L], [-Eq \quad +Eq], [-L \quad +Eq]$, or $[-Eq \quad +L]$

$$\text{Type M force: } \begin{bmatrix} +\text{A} & +\text{L} \\ & -\text{A} \end{bmatrix}, \begin{bmatrix} +\text{A} & +\text{Eq} \\ -\text{L} & \end{bmatrix}, \begin{bmatrix} +\text{A} & -\text{A} \\ -\text{Eq} & \end{bmatrix}, \begin{bmatrix} +\text{A} & +\text{L} \\ -\text{Eq} & \end{bmatrix}, \ldots$$

where the positive and negative signs (+ and −) mean an increase and a decrease respectively. If, for example, a firm purchases goods on account, this accounting event has a type B force, [+A +L], since it increases both the inventory asset account and the account-payable account. Thus, the transaction increases both sides of the fundamental accounting model and does so to the full extent of its monetary size. Accounting events of type 1 have a type B or type O force, depending on the signs (+ and −) of the change in the debit and credit accounts. If these dual accounts have the same sign, the accounting events have a type B force; otherwise, they have a type O force. Accounting events of type 2, 3, or 4 can have a type O or type B force if the multiple accounts on the debit or credit side (or both sides) change in the same direction. Otherwise, they have a type M force.

The Debit, Credit, and Size Forces The force of accounting events, whether it is of type B, O, or M, is generated by a set of ingredient forces. These forces can be called "ingredient" in the sense that they are the necessary elements of the force of accounting events. They have the same magnitude and act together in perpendicular directions under the double entry principle. One of them acts on the debit side of the events, and the other on the credit side. Hereafter, the former is called the debit force of accounting evets, and the latter the credit force. The debit force comes from the debit element (d) of the events, and the credit force from the credit element (c). As a pair, the dual forces act consistently with the duality and balance features of accounting events. And they do so along with another force—the force of the monetary size of the events, which is referred to as the size force of accounting events hereafter. This force is also an ingredient force in that it is a necessary part of the force of accounting events. For a given accounting event, the three ingredient forces have the same magnitude and act together in perpendicular directions under the double entry principle.

In a geometric space, a force is often shown as an arrow. So are the various forces of accounting events. Figure 5.2 shows such arrows in the two-dimensional (debit-credit) Cartesian space. On each coordinate, there is a pair of arrows, starting at the point of origin but pointing in opposite directions. One of them lies on the positive side (+) of the debit coordinate, and the other on the negative side of it. The former represents the debit force of accounting events that increases the debit account. And the latter represents the debit force of accounting events that decreases the debit account. The pair of solid-line arrows on the credit coordinate are interpreted similarly. For a given accounting event, the two adjacent debit and credit force arrows have the same length and point in perpendicular directions. Since there is no third coordinate in the two-dimensional space, the size force arrow of accounting events is not shown in the figure.

In the three-dimensional (debit-credit-size) Cartesian space, there is a third (size) coordinate, which is reserved for the size force of accounting events. This force is

Fig. 5.2 The various forces of accounting events in the two-dimensional (debit-credit) space

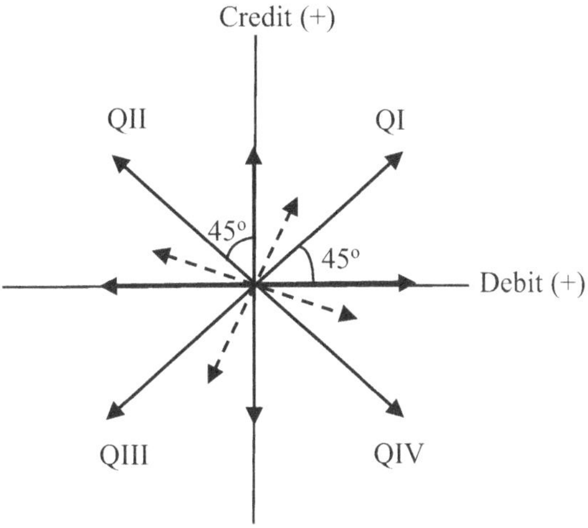

shown by an arrow on the size coordinate. No accounting event can take place with a zero or negative monetary volume. Thus, the size force arrow lies on the positive side of the size coordinate and points straight up. With the size coordinate in place, each of the triple ingredient forces—debit, credit, and size—of accounting events can be shown on a distinct coordinate in the three-dimensional space. They all start at the same point of origin (zero monetary quantity), have the same magnitude, and act in mutually perpendicular directions under the double entry principle.

The Equiforce Besides the three ingredient forces above, every accounting event has another force. This force is generated jointly by the ingredient forces of the event. Hereafter, it is called the equiforce of accounting events. For a given accounting event, its equiforce is the only force that represents the force of the event. As such, it is the only force that reflects the duality and balance features of the accounting event and the contrasting yet symmetric nature of double entries. None of the ingredient forces does the same. When accounting events were said to have a force of type B, O, or M before in this chapter, it meant their equiforce. With this force as well, every accounting event has four kinds of forces—debit, credit, size, and equiforce. These forces co-exist, co-vary, and act together according to the double entry principle. Hence, if any of them is known or determined, the others are also known or determined.

In Fig. 5.2, the equiforces of accounting events are shown by the solid-line force arrows that lie *off*, not on, the debit and credit coordinates. There is one such arrow in each quadrant of the two-dimensional space. Each of them has a slope of 1 since it has an angle of 45° with the adjacent debit and credit coordinates. It is generated by the pair of the adjacent debit and credit forces having the same magnitude, which act together and perpendicularly under the double entry principle. As the debit and credit force arrows start at the point of origin, so does the equiforce arrow.

Conceptually, the equiforce of an accounting event differs from the ingredient forces of the event. While generated jointly by the latter forces, it is the only force that represents the force of the accounting event and reflects the duality and balance features of the event. In this regard, it is double-sided—one side reflecting the debit force and the other reflecting the credit force. Neither the debit force nor the credit force does the same since it is not double-sided. The size force of an accounting event is generated jointly by the debit and credit forces so that it is double-sided as well. Yet it is considered an ingredient force as it is a necessary element of the equiforce of the accounting event. Reflecting the duality and balance features of accounting events, the equiforce arrows lie *off*, not on, the debit and credit coordinates. They have a slope of 1 as they have an angle of 45° with the adjacent debit and credit coordinates.

Similar to forces in general, the various forces of accounting events have both magnitude and direction, and their magnitude is independent of their direction. The same is true for their equiforce as well. For a given accounting event, therefore, the magnitude of its equiforce is the same, regardless of whether the equiforce acts northeast or southwest. This feature of the equiforce is consistent with the notion that the force of one dollar or a one-dollar change is the same, regardless of whether it acts on the debit side or on the credit side.

Having such features, geometrically, the equiforce arrow of accounting events is considered equivalent to the double entry (DE) line—the diagonal of the DE square—in Sect. 3.4, Chap. 3. It divides the DE square into two symmetric right triangles of equal size. Given this, the magnitude of the equiforce of accounting events can be computed according to the well-known Pythagorean theorem as

$$|\mathrm{F_E}| = \left(|\mathrm{F_D}|^2 + |\mathrm{F_C}|^2\right)^{1/2}$$

where $\mathrm{F_E}$, $\mathrm{F_D}$, and $\mathrm{F_C}$ denote the equiforce, debit force, and credit force respectively of accounting events. $|\mathrm{F_E}|$ is equal to the length of their equiforce arrow, namely the length of the DE line.

The discussions above of the forces of accounting events in the two-dimensional space also hold in the three-dimensional (debit-credit-size) space. In this space, the equiforce of accounting events is generated by triple ingredient forces—debit, credit, and size. These forces have the same magnitude and act together in perpendicular directions along the respective coordinates of the three-dimensional space. As such, they define a cubic space—a DE cube. And the diagonal of this cube is equal to the equiforce arrow of accounting events. It has a slope of 0.7071 as it has an angle of 35.2645° with the bottom face of the DE cube. It is equivalent to the double entry (DE) line—the diagonal of the DE cube—in Sect. 4.4, Chap. 4. The magnitude of the equiforce of accounting events is equal to the length of their equiforce arrow. It is computed according to the Pythagorean theorem as

$$|F_E| = \left(|F_D|^2 + |F_C|^2 + |F_S|^2\right)^{1/2}$$

where F_S denotes the force of the size element of the accounting events.

The equiforce arrows shown in Fig. 5.2 before concern accounting events of type 1, having no component transactions. The equiforce arrow of such accounting events can lie in any of the four quadrants (QI, QII, QIII, and QIV) of the two-dimensional space as those accounting events have a type B or type O equiforce. The geometric location of the equforce arrow depends on the direction of the related debit and credit forces. The same is also true in the three-dimensional space; the equiforce arrow of type 1 accounting events can lie in any octant (OI, OII, OIII, or OIV) of the three-dimensional space. Again, this is because type 1 accounting events have an equiforce of type B or type O.

If an accounting event is of type 2, 3, or 4, it has component transactions. Each component transaction has an equiforce of its own, which is of either type B or type O. As such, the equiforce of the accounting event is represented not by a single force arrow but by a set of discrete force arrows. These discrete force arrows may or may not point in the same direction as the accounting event may or may not change the multiple accounts on the debit or credit side (or both sides) in the same direction. If they do, the accounting event has a type B or type O equiforce, although this force is not one continuous force but a multitude of discrete forces. And these discrete forces act in the same direction, and the discrete force arrows are collinear and lie in the same quadrant or octant of the multidimensional space. If the discrete force arrows do not point in the same direction, the accounting event has a type M equiforce, which is represented by a set of discrete equiforce arrows. And these discrete arrows are not collinear. Nor do they lie in the same quadrant or octant of the multidimensional space.

The Interaction Forces There is still another kind of force for accounting events. This force is referred to as the interaction force of accounting events hereafter. Like the debit or credit force, it is also an ingredient force in that it is a necessary part of the force of accounting events. However, it exists not for type 1 accounting events but only for type 2, 3, or 4 accounting events, which have component transactions. It is generated by the interaction of the debit (credit) force of one component transaction and the credit (debit) force of another. For such accounting events, their component transactions are not considered independent of each other but interrelated since those component transactions are executed in bundle when the accounting events are executed. As such, there exists a *cross* interaction of the debit force of one component transaction and the credit force of another, or vice versa. This cross interaction generates a force—an interaction force. By nature, type 1 accounting events cannot have an interaction force since they have no component transactions. When type 2, 3, or 4 accounting events have interaction forces, they have two of such forces at a minimum because, when they have component transactions, they have a minimum of two component transactions.

Since an interaction force of accounting events is generated by the cross interaction of the debit (credit) force of one component transaction and the credit (debit) force of another, it does not reflect the duality and balance features of the accounting event or a particular component transaction. Nor does it reflect the contrasting yet symmetric nature of double entries.

As with forces in general, the interaction forces of accounting events can also be shown with arrows. They are shown by the dashed-line arrows in Fig. 5.2 before. Unlike the debit and credit force arrows, they lie *off*, not on, the debit and credit coordinates. Also, unlike the equiforce arrows, they may or may not have an angle of 45° with the adjacent debit and credit coordinates since they do not need to reflect the duality and balance features of component transactions or the symmetric nature of double entries.

While the geometric space represented by an equiforce is a square or cube, the geometric space represented by the interaction forces of component transactions may or may not be a square or cube. It can be a rectangle in the two-dimensional space or a rectangular cuboid in the three-dimensional space if the component transactions of an accounting event do not have the same monetary size. Despite the differences, the magnitude of an interaction force is measured and shown in the same way the magnitude of the equiforce of the accounting event is. That is, it is measured and shown by the length of the diagonal of the square or rectangle, or the diagonal of the cube or rectangular cuboid, defined by the debit (credit) of one component transaction and the credit (debit) of another.

Nonnegativity of the Forces

A force has a direction. If a force acts in the positive (negative) direction of a point of reference or the point of origin, it is called a positive (negative) force. As such, whether a force is positive or negative is a matter of its direction relative to the point of origin or reference. As such, it is a matter of designation, not substance. Force is not like temperature, which can be intrinsically positive (hot) or negative (cold). Nor is it like color, which can be intrinsically different (e.g., black or white). Rather, it is more like distance, size, or space, which is not intrinsically positive or negative. When a force is shown with a positive or negative sign (+ or −) or with an arrow pointing to the left or to the right, it is a matter of designation or presentation. It does not mean that the force is intrinsically positive or negative. Nor does it mean that the magnitude of the force varies with its direction, positive or negative. The magnitude of a force is the same, regardless of whether the underlying force acts in the positive or negative direction from the point of reference.

The same is also true for the forces of accounting events. They are nonnegative in substance, whether they are the debit or credit force, or the equiforce or interaction force. Also, as with the magnitude of forces in general, the magnitude of the forces of accounting events is independent of their directions. It is the same, regardless of whether the forces act on the debit or credit side or point northeast or southwest. What this means is that the magnitude of the force of one dollar or a one-dollar change caused by an accounting event is the same, regardless of whether the force acts on the debit side or credit side of the event. It is also the same, regardless of whether the accounting event causing the one-dollar change is of type 1 or type 4, all else equal (e.g., the monetary size of the event).

Geometrically, the forces of accounting events can point in a positive or negative direction or in different directions in the two- or three-dimensional Cartesian space. If a force of an accounting event increases an account on its debit (credit), it is shown on the positive side of the debit (credit) coordinate. And vice versa. Again, this is a matter of presentation, not substance. The forces of accounting events are not like temperature or color, which can be intrinsically different; they are more like distance or space, or even money, which cannot be intrinsically positive or negative. There is no such thing as an intrinsically positive or negative debit or credit force, or an intrinsically positive or negative equiforce, of accounting events.

The nonnegativity of the forces of accounting events is consistent with the nonnegativity-of-monetary convention in accounting (see Sect. 2.2, Chap. 2). The forces of accounting events are always considered positive since the debit, credit, and size elements of the events cannot be intrinsically negative, nor even be zero. No accounting event can take place with a negative or zero debit, credit, or size (monetary volume).

5.2 The Power of Accounting Events

The Power of Accounting Events and Its Duality

Power is generated by a force, not vice versa. Hence, if anything has a force, it has power. Although interrelated, the two are not the same concept. While force has magnitude and direction, power has magnitude but no direction. It is a directionless scalar quantity and independent of the direction of the underlying force. For a given force, therefore, the power is the same, regardless of whether the force acts to the left or to the right.

As with force, power has different meanings, depending on how it is defined or seen. Conceptually, power means the strength of a force to affect something (tangible or intangible) or make something change, happen, or move. As with force, power can be physical. If a force is concerned with an object or something physical, it is a physical property and the related power is also physical. And the power means the strength of the underlying force to make the object change or move. Examples of such power are the power of a horse—the strength of the horse to pull the weight of a physical object—and the power of a car engine—the strength of the engine to speed up the car per unit of time. If a force is a nonphysical property, the related power is also nonphysical. It means the strength of a nonphysical force to have an impact on something. Examples of such power are the power of people's force to change a political or social policy and the power of a police force to enforce laws. The power of Nature is another such example as the force of Nature is also a nonphysical force.

As accounting events have a force—an equiforce in particular, they have power. This power comes from the equiforce of the events, just as power in general comes from its underlying force. Since the equiforce of accounting events comes from their ingredient forces—debit, credit, and size, the power of accounting events originates from those ingredient forces. The power of accounting events is nonphysical since the events are nonphysical. For an accounting event, its power means the strength of its equiforce or the extent of its impact on the stock balance of the sources and uses of the capital resources of a business entity (firm) and their composition and relationship as shown by the fundamental accounting model. As with power in general, the power of accounting events is a scalar quantity. It has magnitude but no direction. The magnitude of the power of accounting events is independent of the direction of their equiforce or ingredient forces. Thus, for a given accounting event, its power is the same, regardless of whether its equiforce acts northeast or southwest, that is, whether the accounting event increases or decreases the debit and credit accounts.

As the equiforce of accounting events reflects the duality and balance features of the events, so does their power. What this means is that there exists a balance of power between the debit force and credit force of accounting events. And the equiforce and power of accounting events act or work accordingly. When an accounting event takes place, the power of a one-dollar change (increase or decrease) on the debit side of the event is equal to the power of the one-dollar change (increase or decrease) on the credit side. Also, the power of a one-dollar change is the same, whether it comes from the force of a simple accounting event or a complex accounting event, all else equal. In other words, the balance-of-power notion works, regardless of whether the power of a one-dollar change (increase or decrease) is caused by the type B or O equiforce, or the type M equiforce, of accounting events when all else is the same (e.g., the monetary size of the events).

Determinants of the Power of Accounting Events

How the power of accounting events is determined may be explained based on a well-known theory on power. In physics, for example, power is known to be determined by three factors—force, distance, and time.[3] The magnitude of power is increasing with the force and distance factors but decreasing with the time factor. It is measured by the extent of an object moved, or the speed of a task performed, by the underlying force. As a nondirectional scalar quantity, the magnitude of power is positively related to the magnitude of its underlying force and the distance of the object moved or the extent of the task accomplished by the force. But it is negatively related to the time taken by the force to do so. The greater the magnitude of the force, and the longer the distance of the object moved, the greater is the magnitude of power, all else equal. And the shorter the time taken by the force to do so, the greater is the magnitude of power. The theory on the determinant of power does not show whether the direction of the force and the location of the object moved are also determining factors. The reason appears to be because power is an absolute scalar quantity so that it is independent of the direction of its underlying force and the location where the force acts.

[3] See https://en.wikipedia.org/wiki/Power(physics) (August 2020).

Although the theory on power above pertains to the power of physical objects or things in general, it offers an insight into how the power of accounting events as a nonphysical property is determined. Given the theory, it can be said that the power of accounting events depends on their equiforce, which is generated by a set of ingredient forces—debit, credit, and size. It is increasing with the magnitude of the equiforce, namely the magnitude of the ingredient forces of accounting events. The "distance" and "time" factors are not considered relevant to the power of accounting events since neither the time taken to execute accounting events nor the location where accounting events take place matters to the power of accounting events. This statement is reasonable, given that neither of these two factors is formally recognized when accounting events are recorded under the double entry principle. Consistently, neither the distance factor nor the time factor is embedded in the fundamental accounting model. When it comes to the power of accounting events, therefore, what really matters is the force factor—their equiforce in particular. And the magnitude of this equiforce depends on the (monetary) size of accounting events.

What the above means is that, when it comes to the power of accounting events, their size matters. This statement is consistent with the general notion that, when it comes to power, size matters. Thus, different accounting events can have the same power if they have the same monetary size, all else equal, even if they differ in nature or complexity. Thus, it can be said that the power of an accounting event is monotonically increasing with the magnitude of its equiforce, namely the magnitude of the ingredient forces of the accounting event. As these ingredient forces depend on the (monetary) size of the accounting event, so does the power of the event. This is so, regardless of whether accounting events are of type 1 or type 4, or whether their equiforce is of type B, O, or M. A difference is that, if an accounting event is of type 1, having no component transactions, its equiforce is one continuous force. Thus, the power of the event has one continuous magnitude.

If an accounting event is of type 2, 3, or 4, having component transactions, its equiforce is divided into a multitude of discrete equiforces. Thus, the power of the accounting event has a multitude of discrete magnitudes. It is equal to the power of the component transactions combined plus the power of their interactions combined. It is increasing with the magnitude of the interaction forces, as well as the equiforces, of the component transactions. Given this, it can be said that two accounting events having the same monetary size have the same power, all else equal, even if they differ in nature or complexity. This is consistent with the balance-of-power notion described before. This notion holds, regardless of whether the component equiforces or interaction forces act in the same direction or in different directions. It does so because power is a directionless scalar quantity and the balance-of-power notion works for all types of accounting events. Again, what this means is that what matters to the power of accounting events is their size, not complexity or simplicity.

Measuring the Power of Accounting Events

The power of an object or a thing, whether tangible or intangible, is measured by the extent of the object changed or moved or the amount of the work done by the underlying force, which is expressed in a measurement unit. For example, the power of a car engine is measured by the distance in miles the car can run per unit of time at its full speed. And the power of a horse is measured by the weights in pounds of an object or a thing the horse can pull with a maximum force. These are examples of the measurement of power as a physical property. If power is a nonphysical property like the power of people, measuring the power is not as simple. Yet such power can still be measured by the extent or outcome of the impact of the underlying force on something. For example, the power of people can be measured by the extent to which they change a political system or a social policy with their force.

The power of accounting events as a nonphysical property can also be measured by the extent of their impact on something. Specifically, it can be measured by the extent of their impact on the sources and uses of the capital resources of a business entity as shown by the fundamental accounting model. Or it can be measured by the extent of their impact on the debit and credit accounts directly affected.

Numerical Measurement Since the forces of accounting events are nonphysical, measuring their power is tricky. One way to measure the power of accounting events is to assess the extent of their impact on the debit and credit accounts directly affected *and* the elements (assets, liabilities, and owner equity) of the fundamental accounting model. This is reasonable, given that an accounting event is defined as an event that affects the model in one way or another, regardless of whether it is an arm's-length transaction or not (see Sect. 1.1, Chap. 1). Specifically, the power of an accounting event is measured by the extent to which the event changes the debit and credit accounts directly affected and the relationships of these accounts with other accounts as part of the elements of the fundamental accounting model. The greater the extent of the change, the larger the impact of the accounting event, and the greater is the power of the event, when all else is equal.

To illustrate, consider a firm that has the sources and uses of its capital resources (assets, liabilities, and owner equity) as shown by the fundamental accounting model below.

Cash \$20 + Inventory (Inv) \$10 = Liabilities (Liab) \$10 + Owner Equity (Eq) \$20

Thus, the elements of the model have the following relationships:

$$\text{Cash}/\text{Inv} \rightarrow 2 \text{ to } 1 \rightarrow \text{Cash} = 200\% \text{ of Inv}$$

$$\text{Cash}/\text{Eq} \rightarrow 1 \text{ to } 1 \rightarrow \text{Cash} = 100\% \text{ of Eq}$$

$$\text{Inv}/\text{Eq} \rightarrow 1 \text{ to } 2 \rightarrow \text{Inv} = 50\% \text{ of Eq}$$

$$\text{Liab}/\text{Eq} \rightarrow 1 \text{ to } 2 \rightarrow \text{Liab} = 50\% \text{ of Eq}$$

$$\text{TA}/\text{Liab} \rightarrow 3 \text{ to } 1 \rightarrow \text{TA} = 300\% \text{ of Liab}$$

where TA denotes total assets.

Suppose now that the firm purchases goods for $10 in cash. This accounting event is of type 1 as it has no component transaction. It increases the inventory asset as a debit and decreases the cash as a credit. Thus, it only affects the debit side of the fundamental accounting model and has an equiforce of type O. As the event has no component transaction, it has no interaction forces of component transactions. And the equiforce of the event has one continuous magnitude. So does the power of the event.

Shown below is the impact of the accounting event on the two accounts directly affected and on the relationships among the elements of the fundamental accounting model.

$$\text{Cash } \$10 + \text{Inv } \$20 = \text{Liab } \$10 + \text{Eq } \$20$$

$$\text{Cash} \rightarrow \ -50\% \text{ points}$$

$$\text{Inv} \rightarrow \ +100\% \text{ points}$$

$$\text{Cash/Inv} \rightarrow \text{Cash} = 50\% \text{ of Inv} \rightarrow \ -150\% \text{ points}$$

$$\text{Cash/Eq} \rightarrow \text{Cash} = 50\% \text{ of Eq} \rightarrow \ -50\% \text{ points}$$

$$\text{Inv/Eq} \rightarrow \text{Inv} = 100\% \text{ of Eq} \rightarrow \ +50\% \text{ points}$$

$$\text{Liab/Eq} \rightarrow \text{Liab} = 50\% \text{ of Eq} \rightarrow \text{no change}$$

$$\text{TA/Liab} \rightarrow \text{TA} = 300\% \text{ of Liab} \rightarrow \text{no change}$$

The accounting event decreases the cash account by 50% and increases the inventory asset account by 100%. As a result, the relationships among the elements of the fundamental accounting model also change as shown. For example, the cash-to-inventory asset ratio changes from 2:1 to 1:2, decreasing by 150 percentage points. And the cash-to-equity ratio changes from 1:1 to 1:2, decreasing by 50 percentage points.

Given all the changes, the power of the accounting event can be measured by a simple average of the percentage-point changes (in absolute terms) in the two accounts directly affected (75%) or by a weighted average of those percentage-point changes (66.7%). Or it may be measured by a simple average of the percentage-point changes in the two accounts directly affected *and* in the relationships among all the accounts (80%). Or it may be measured by the spread of those percentage-point changes (250%).[4] In each case, the larger the percentage-point change or the spread of those changes, the greater is the power of the accounting event.

---

[4]Those percentage-point changes are computed as follows (in %): (100 + |−50|) / 2 = 75, [|−50|(0.20) + (100)(0.10)] / (20 + 10) = 66.7, (|−50| + 100 + |−150| + |−50| + 50) / 5 = 80, and 100 − |−150| = 250.

The percentage-point change-based measure of power is a rough measure yet has a merit; it considers not only the impact of an accounting event on the debit and credit accounts directly affected but also the impact of the event on the relationships among all the accounts, including those that are not directly affected by the event. Yet it has a drawback; the percentage-point change-based measure of power varies not only with the (monetary) size of accounting events but also with the *base* monetary balance (extant balance) of the accounts (e.g., cash and inventory asset) directly affected by the events; the smaller this base balance, the larger the percentage-point changes, and thus the greater is the impact, hence power, of the accounting events even if all else is equal. This is not consistent with the notion that what matters to the power of accounting events is their (monetary) size, not the extant balance of the accounts directly affected. The power of two accounting events can differ simply because the extant balance of the debit and credit accounts affected differs, even if the two events have the same monetary size. Also, the percentage-point change-based measure of power is inconsistent with the balance-of-power notion indicated before. The power of an accounting event causing a one-dollar change can differ from the power of another accounting event causing the same one-dollar change, simply because the extant balance of the debit and credit accounts directly affected differs between the two events.

Geometric Measurement The power of accounting events can also be measured geometrically in a multidimensional space. It can be measured by the size of the geometric space determined by the forces of the accounting events. This space is the DE square of the accounting events in the two-dimensional (debit-credit) space and the DE cube in the three-dimensional (debit-credit-size) space. Although the two, power and geometric space, are not the same concept, their measures in absolute terms can be the same. A rationale for the space-based measurement of power is that an accounting event has a capability of stretching its equiforce along with its ingredient forces, thereby creating a geometric space of some size in the two- and three-dimensional spaces respectively. The larger the monetary size of accounting events, the farther their forces are stretched out, the larger the DE space created, and thus the greater is the power of the accounting event, all else equal. This statement is consistent with the notion that when it comes to the power of accounting events, their monetary size matters or, more generally, that when it comes to power, size matters.

Once the DE square or DE cube of an accounting event is determined, the power of the event may be measured in terms of the length of the diagonal (DE line) of the DE square or DE cube, the distance of the diagonal vertex (DE point) from the point of origin, or the size of the DE square or DE cube. While the line- and point-based measurement of power may be plausible, the space-based measurement of power is conceptually more reasonable since it is more consistent with the nature of power in general. Unlike force, which is a linear and additive concept, power is a nonlinear and multiplicative concept. In mathematics, for example, the term *power* means something multiplicative or exponential, rather than additive or linear. Specifically, it means the exponent of a number—the number of repeated multiplications of a base number. If, for example, the power of 5 as a base number is 3, this means that the base number 5 is multiplied three times: $5^3 = 5 \times 5 \times 5 = 125$, not 5×3 or $5 + 5 + 5 = 15$. Thus, the power 3 of the number 5 means 125, not 15. What this means in view of force and power is that, if a force increases three times, the related power increases not just three times but exponentially.

Given the nature of power as a multiplicative concept, the power of accounting events can also be considered multiplicative. Thus, it can be measured in terms of the size of the DE square or DE cube of the events, which is also multiplicative in nature. Thus, the space-based measurement of the power of accounting events is conceptually more reasonable than the line- and point-based measurement. In other words, measuring the power of an accounting event by the product d.c. or d.c.q is more reasonable than measuring it by the simple sum: |d| + |c| or |d| + |c| + q. The product d.c. equals the size of the DE square and d.c.q equals the size of the DE cube. Thus, if an accounting event is 10 times larger in monetary size than another accounting event, the former is 100 times, not 10 times, more powerful than the latter.

When the power of an accounting event is measured as said above, it is equal in magnitude to the space-based measure of the size of the accounting event, namely the size of the DE square in Chap. 3 and the DE cube in Chap. 4. Although the two, power and size, are not the same concept, their measures as scalar numbers can be the same. For accounting events, the former is one of their qualitative (nonmonetary) properties, while the latter is one of their intrinsic elements (debit, credit, and size). Thus, the two are interrelated but are not the same in nature.

If an accounting event is of type 1, the power of the event measured by the size of its DE space (square or cube) is one continuous magnitude as the event has no component transactions. If an accounting event is of type 2, 3, or 4, the power of the event is not one continuous magnitude but a multitude of discrete magnitudes as the event has component transactions. For such accounting events, the power is measured by the size of the multiple component spaces (squares and rectangles, or cubes and rectangular cuboids) combined. It includes not only the direct power of the component transactions combined but also their interaction power. The former is the power generated by the debit and credit forces of each component transaction, and the latter is the power generated by the interaction forces of different component transactions. Measuring the power of such accounting events as said is consistent with the notion of aggregation, which is a linearly additive concept. It is also consistent with the "whole-parts" relation—that the "whole" is the same as the sum of its "parts."

The space-based measurement of the power of accounting events, relative to the percentage-point change-based measurement discussed before, has some merits. First of all, it is consistent with the nature of power as a nonlinear, multiplicative concept. It is also consistent with the notion that size matters when it comes to power and that what matters to the power of accounting events is their monetary size, not complexity or simplicity. It is also consistent with the balance-of-power notion—that the force of a one-dollar change has the same power, whether it acts on the debit or credit side or whether it is caused by an accounting event of type 1 or type 4, all else equal. More importantly, the space-based measurement of the power of accounting events is not affected by the base (extant) balance of the debit and credit accounts directly affected by the events.

While having such merits, the space-based measurement of power has a drawback. Unlike the percentage-point change-based measure of power before, it only reflects the impact of accounting events on the debit and credit accounts directly affected by the events. It does not reflect the impact of the accounting events on the relationship of those accounts with other accounts that are not affected by the events yet are still part of the elements of the fundamental accounting model.

5.3 The Forces and Power of Accounting Events in Geometric Settings

Illustrations in the Two-Dimensional Space

The geometric presentation and measurement of the forces and power of accounting events discussed thus far can be illustrated with specific transaction examples. This section shows it with two transaction examples from Chap. 3. As for the power of accounting events, the illustrations focus on the space-based presentation and measurement of power since this is more consistent with the nature of power than the line- or point-based presentation and measurement of power as indicated before.

Illustration 1 Consider the type 1 transaction example—that a firm sells goods for \$100 in cash. This event has no component transactions so that it has a type B equiforce, since it increases the debit and credit accounts alike to the full extent of its monetary size (\$100). Figure 5.3, adapted from Fig. 3.3, presents the forces and power of the accounting event in the two-dimensional (debit-credit) geometric space. The shaded square is the DE square of the event. The arrows on the debit and credit coordinates represent the debit and credit forces of the accounting event, and the diagonal arrow represents the equiforce. As the accounting event increases the debit and credit accounts alike, the equiforce arrow lies in QI of the two-dimensional space and points northeast. It is determined jointly by the pair of debit and credit forces that

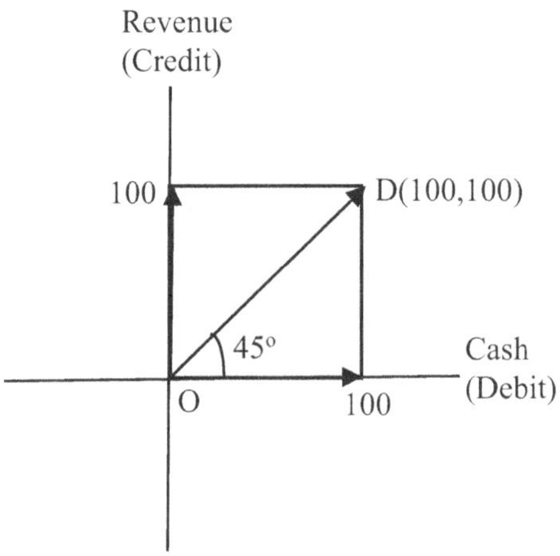

Fig. 5.3 Presenting the forces and power of a type 1 accounting event in the two-dimensional (debit-credit) space

have the same magnitude (100) and act in perpendicular directions under the double entry principle. It has a slope of 1 as it has an angle of 45° with the debit and credit coordinates.

The magnitude of the debit and credit forces of the accounting event is equal to the length of the respective debit and credit force arrows, which is 100. And the magnitude of the equiforce of the accounting event is measured by the length of the equiforce arrow according to the Pythagorean theorem as

$$|F_E| = \left[100^2 + 100^2\right]^{1/2} = 141.4214$$

In magnitude, $|F_E|$ is equal to the line-based measure (Q_L) of the size of the accounting event in Sect. 3.4, Chap. 3. Although force and size are not the same concept, their measures can be the same. $|F_E|$ is concerned with the force of the accounting event as one of its qualitative properties, while Q_L is related to the size of the accounting event as one of its intrinsic elements.

The power of the accounting event, if measured geometrically, is equal to the size of the DE square of the event. It is computed as.

$$P = d.c = (100)(100) = 100^2 au^2$$

where "au" denotes an "area unit," the same measurement unit used to denote the size of accounting events in Chap. 3. As a scalar quantity, P is considered a geometric measure of the extent of the impact of the accounting event on the two accounts, cash and revenue, directly affected. It is a geometric measure of the strength of the capability of the accounting event to stretch its equiforce arrow along with the related debit and credit force arrows, thereby creating a geometric square that has a size of 100^2 au^2. In magnitude, this measure is the same as the size of the DE square of the event in Chap. 3. It is also considered a geometric measure of the impact of the accounting event on the elements of the fundamental accounting model, namely the sources and uses of the capital resources of the firm. As the event is of type 1, its equiforce has one continuous magnitude. So does its power.

Illustration 2 Consider the type 4 accounting event—that a firm recognizes an interest expense of $330 on a bond liability, pays the cash $350 for the accrued interest on the bond liability, and amortizes a bond discount of $100. The interest paid in cash includes a previously accrued but unpaid interest of $120 on the bond liability. The accounting event has three component transactions, and the multiple accounts on each side of the event do not change in the same direction. As such, it affects both sides of the fundamental accounting model but not to the full extent of its monetary size $450. Hence, the accounting event has a type M equiforce. As the event has three component transactions, this equiforce is divided into multiple discrete forces.

Figure 5.4, adapted from Fig. 3.10, presents the forces and power of the accounting event in the two-dimensional space. The large unshaded square is the DE square of the event, and each small shaded DE square represents a component transaction.

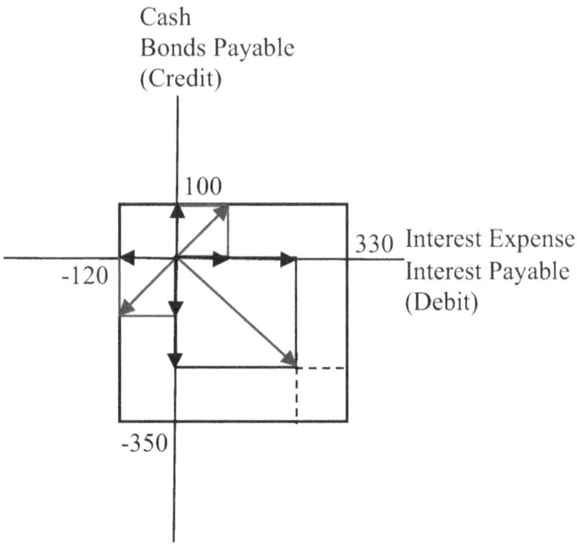

Fig. 5.4 Presenting the forces and power of a type 4 accounting event in the two-dimensional (debit-credit) space

Given the three component transactions, the DE square of the event is divided into three component squares and six rectangles. And the accounting event has three discrete equiforces, one for each component transaction, and six interaction forces. The diagonal arrow of each small square represents the equiforce of a component transaction (the interaction force arrows are not shown to keep the figure simple). Each rectangle is determined by the cross interaction of the debit (credit) force of one component transaction and the credit (debit) force of another. For example, the unshaded rectangle in QI of the two-dimensional space is determined by the interaction of the debit force of the interest expense \$230 in one component transaction and the credit force of the bond liability (discount amortization) \$100 in another component transaction.

The magnitude of the debit and credit forces of each component transaction is measured by the length of the respective force arrows on the debit and credit coordinates. The magnitude of the equiforce of each component transaction is measured by the length of its equiforce arrow, which is equal to the diagonal of the related small DE square. And the magnitude of the equiforce of the accounting event as a whole is measured by the length of the equiforce arrows of the three component transactions combined. It is computed according to the Pythagorean theorem as.

$$|F_E| = \left(100^2 + 100^2\right)^{1/2} + \left(|-120|^2 + |-120|^2\right)^{1/2} + \left(230^2 + |-230|^2\right)^{1/2}$$
$$= 636.3961$$

As seen, $|F_E|$ is not one continuous magnitude but a multitude of discrete magnitudes. Measuring the magnitude of the equiforce of the accounting event as such a multitude is consistent with the notion of linear aggregation and the "whole-parts" relation indicated before. In magnitude, $|F_E|$ equals the line-based measure (Q_L) of the size of the accounting event seen in Sect. 3.4, Chap. 3. Although the two, force and size, are not the same concept, their measures in absolute terms can be the same.

Since the accounting event is represented by three small squares and six rectangles, its power is measured by the size of these squares and rectangles combined. It is computed as.[5]

$$\begin{aligned} P &= (100 + 230 + |-120|)(100 + |-120| + |-230|) \\ &= 450^2 au^2 \end{aligned}$$

The space-based measure P of power is equal in magnitude to the space-based measure (Q_S) of the size of the accounting event in Chap. 3. Since the event has three component transactions, P is not one continuous magnitude but a multitude of discrete magnitudes. Again, measuring P as such a multitude poses no problem as power is a directionless scalar quantity. It is consistent with the "whole-parts" relation and the notion of linear aggregation. The power of this accounting event differs from the power of the type 1 accounting event before because its size differs. This event is 4.5 times larger in size. So is its equiforce. As such, the power of the event is 20.25 times greater as $4.5^2 = 20.25$. This is because power is a multiplicative concept and increases exponentially as the underlying force increases.

Illustrations in the Three-Dimensional Space

The geometric presentation and measurement of the forces and power of accounting events can also be illustrated in the three-dimensional (debit-credit-size) setting as they can be in the two-dimensional case.

Illustration 1 Consider again the type 1 transaction example before. Figure 5.5, adapted from Fig. 4.3, graphically shows the forces and power of the accounting event in the three-dimensional (debit-credit-size) space. The arrows on the three coordinates represent the debit, credit, and size forces respectively of the event. This accounting event affects the quantity of both sides of the fundamental accounting model to the full extent of its monetary size. Thus, the event has a type B equiforce. This force is shown by the diagonal arrow OD of the shaded DE cube.

The magnitude of the debit, credit, and size forces respectively of the accounting event is 100, equal to the length of the respective force arrows shown in the figure. And the magnitude of the equiforce of the accounting event is equal to the length of the diagonal arrow. It is computed according to the Pythagorean theorem as.

[5]To see it algebraically, the expression of P can be expanded as 100.100 + 100.|−120| + 100.|−230| + 230.100 + 230.|−120| + 230.|−230| + |−120|.100 + |−120|.|−120| + |−120|.|−230|. There are three direct products 100.100, 230.|−230|, and |−120|.|−120|, each being equal to the size of the related DE square. And there are six cross products, each being equal to the size of the related rectangle.

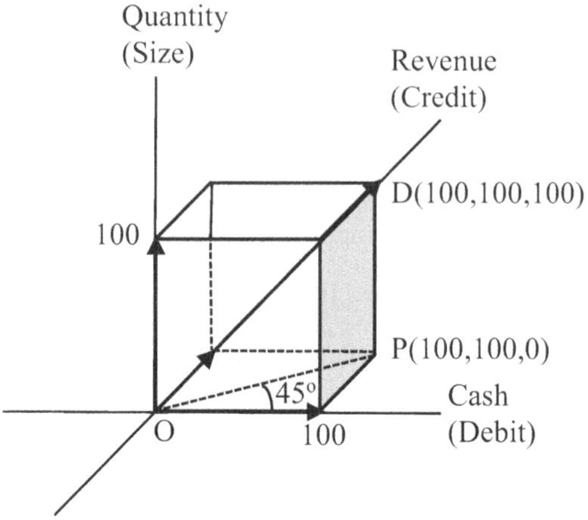

Fig. 5.5 Presenting the forces and power of a type 1 accounting event in the three-dimensional (debit-credit-size) space

$$|F_E| = \left[100^2 + 100^2 + 100^2\right]^{1/2} = 173.2051$$

The magnitude $|F_E|$ of the equiforce of the event is equal to the line-based measure (Q_L) of the size of the event seen in Sect. 4.4, Chap. 4. Although force and size are not the same concept, their measures as scalar numbers are the same.

When measured geometrically, the power of the accounting event is equal to the size of the DE cube. It is computed as

$$P = d.c.q = (100)(100)(100) = 100^3 vu^3$$

where "vu" denotes a "volume unit," the same measurement unit used to denote the size (volume) of accounting events in Chap. 4. P is equal in magnitude to the space-based measure (Q_S) of the size (quantity) of the accounting event there. As a scalar quantity, P is a geometric measure of the impact of the accounting event on the two accounts (cash and revenue) directly affected by the event. That is, it is a geometric measure of the strength of the capability of the accounting event to stretch the equiforce arrow along with the debit, credit, and size force arrows, thereby generating a cubic space that has a size of 100^3 vu^3. As the accounting event is of type 1, its equiforce is one continuous magnitude. So is its power.

Illustration 2 Consider the type 4 accounting event—that a firm recognizes an interest expense, amortizes a bond discount, and pays an accrued but unpaid interest, all relating to a bond liability that was issued at a discount some time ago. This accounting event has three component transactions and two accounts on each side, debit and credit. And the two accounts on each side do not change in the same direction, one increasing and the other decreasing. The accounting event affects both sides of the fundamental accounting model, but not to the full extent of its monetary size. Thus, it has a type M equiforce. Figure 5.6, adapted from Fig. 4.9, graphically presents the forces and power of the accounting event. The large unshaded cube is the DE cube of the event. As the event has three component transactions, there are

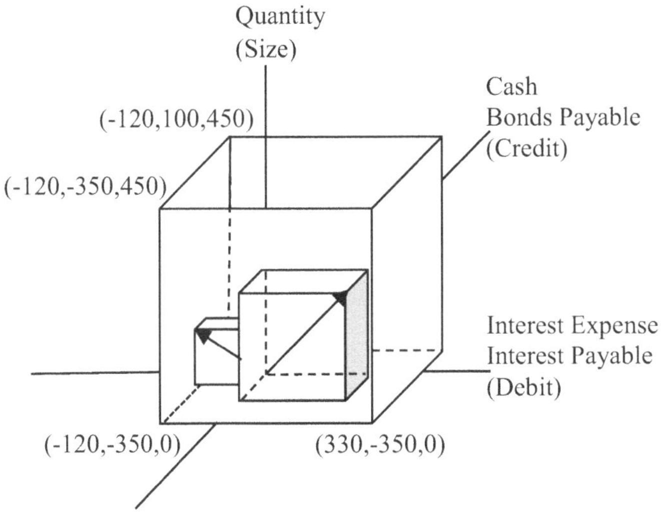

Fig. 5.6 Presenting the forces and power of a type 4 accounting event in the three-dimensional (debit-credit-size) space

three small DE cubes—one for each component transaction—and 24 rectangular cuboids—one for each cross interaction of the debit and credit forces of the three component transactions.[6] And the diagonal arrow of each small DE cube represents the equiforce of the related component transaction. The small cube in OIII represents the payment of the interest payable $120. And the small cube in OIV represents the accrual of the interest expense $330, which includes the amortization of the bond discount $100.

Given the three component transactions, the debit, credit, and size forces respectively of the accounting event are divided into three discrete forces (although the related force arrows are not shown to keep the figure simple). The magnitude of each of those forces is equal to the magnitude of the related discrete forces combined. As the multiple accounts on the debit and credit sides respectively do not change in the same direction, the equiforces of the three component transactions do not act in the same direction and the accounting event has a type M equiforce. This force is represented not by a single equiforce arrow but by a trio of noncollinear equiforce arrows. Although these force arrows are not collinear, their magnitudes can be added to get the magnitude of the equiforce of the accounting event as shown below.

$$
\begin{aligned}
|F_E| &= \left(100^2 + 100^2 + 100^2\right)^{1/2} + \left(|-120|^2 + |-120|^2 + 120^2\right)^{1/2} \\
&\quad + \left(230^2 + |-230|^2 + 230^2\right)^{1/2} \\
&= 779.4229
\end{aligned}
$$

[6]The small DE cube of the component transaction that increases the bond liability due to the bond discount amortization ($100) lies in OI of the three-dimensional space (not seen in the figure).

$|F_E|$ is equal in magnitude to the line-based measure (Q_L) of the size of the accounting event seen in Sect. 4.4, Chap. 4. Measuring the magnitude of the equiforce of the accounting event as shown is consistent with the notion that the "whole" is the same as the sum of its "parts."

If measured geometrically, the power of the accounting event equals the size of the three small cubes combined plus the size of the 24 rectangular cuboids combined. It is computed as[7]

$$\begin{aligned} P &= |d||c|q \\ &= (100 + 230 + |-120|)(100 + |-120| + |-230|)(450) \\ &= 450^3 vu^3 \end{aligned}$$

In magnitude, P is equal to the space-based measure (Q_S) of the size of the accounting event in Sect. 4.2, Chap. 4. It is not one continuous magnitude but a multitude of discrete magnitudes as the equiforce of the event is not one continuous force. Computing P as such a multitude poses no problem since the magnitude of power is an absolute scalar quantity. Also, it is consistent with the "whole-parts" relation and with the notion of linear aggregation mentioned earlier. When compared to the type 1 accounting event before, this event is 91.125 times more powerful ($4.5^3 = 91.125$ or $20.25 \times 4.5 = 91.125$) because the equiforce of the event is 4.5 times stronger as its (monetary) size is 4.5 times larger. This is consistent with the notion that size matters when it comes to power and that power is a multiplicative, not an additive, concept and increases exponentially as the underlying force increases.

5.4 The Full vs. Net Force of Accounting Events

As for type 2, 3, or 4 accounting events having component transactions, a question arises as to whether their force and power can be defined and measured based on their *net force* rather than full force. The answer is *no* since the net force of such accounting events is not the same as their full force. Nor is their power generated by their net force the same as their power generated by their full force. Here, the full force of accounting events means the equiforce of the events that affects all debit and credit accounts involved, without one account being offset against the other. And the net force means the equiforce of the accounting events that is defined after one account is offset against the other (when possible). As such, the full force includes not only the equiforces of all component transactions but also all interaction forces as they actually exist. However, the net force of accounting events does not include the

[7] If expanded, the expression of P yields the algebraic terms for the size of the three component DE cubes and the 24 rectangular cuboids respectively.

equiforces of component transactions to their full extent and all interaction forces as they actually exist. That is, it does not reflect the full effect of accounting events on all the debit and credit accounts affected. As a result, the net force of accounting events does not turn into their full power. Besides, a net force is not conceptually consistent with the "whole-parts" relation mentioned before. Nor is the power generated by the net force consistent with this same relation.

The point of the discussion above may be explained with the type 4 transaction example before. If only the net effects of the transaction on the debit and credit accounts affected were considered, the following journal entry would be made to record the transaction:

Owner equity (interest expense)	\$330	
Liability (interest payable)	20	
Asset (cash)		\$350

The decrease in owner equity by \$330 is due to the accrual of the current-period interest expense \$330, and the decrease in liability by \$20 is due to the excess of the interest payable \$120 paid in cash over the increase in the bond liability due to the amortization of the bond discount \$100. And the decrease in total assets by \$350 is due to the payment in cash of the interest \$350. As a result, the accounting event decreases the liability of the firm only by \$20, not \$120. The effect of the accounting event on the bond liability via the amortization of the bond discount is missing. That is, this component transaction is not recognized in full. Further, the monetary size of the accounting event is shrunk to \$350 from \$450. What all this means is that, if the accounting event is recorded as shown, its effect on all the debit and credit accounts affected is not recognized to the full extent of the monetary size of the event. Nor is the equiforce of the event fully revealed as it actually exists. The same is also true for all ingredient forces of the event.

Given the above, the net force of the accounting event does not turn into the full power of the event, which is determined by the full monetary size \$450 of the event. The debit force of the accounting event generated by $(d - |a|)$ is not as large as the debit force generated by $(d + |b|)$. Similarly, the credit force of the accounting event generated by $(c - |b|)$ is not as large as the credit force generated by $(c + |a|)$. Therefore, the power of the accounting event measured by the size of the DE square equal to $(d - |a|)(c - |b|)$ is not as great as the power of the event measured by the size of the DE square equal to $(d + |b|)(c + |a|)$. The same is also true for other accounting events that have component transactions, whether they are of type 2, 3, or 4.

The Double Entry Accounting in Vector Spaces

6

This chapter discusses how double entry accounting can work in vector spaces. It does so by showing how to present accounting transactions (events) in vector spaces yet according to the double entry principle. The geometric settings used to do so are two- and three-dimensional vector spaces, where a point, line, and space are all a vector or a set of vectors. Accounting events have three intrinsic elements—debit, credit, and size, which were denoted as d, c, and q in previous chapters. Thus, they can be presented as a vector yet according to the double entry principle. While it is a convention to write vectors as column vectors in linear algebra, all vectors in this chapter are row vectors (for convenience) unless otherwise indicated.

6.1 Vectors and Vector Spaces for Double Entry Accounting

Vectors and Vector Spaces in General

Conceptually, a vector is considered a quantity that has magnitude and direction. This quantity is a vector quantity, which is distinguished from a scalar quantity. The latter has magnitude but no direction. The magnitude of a vector means its head-to-tail length, and the direction means where its head points. For a given vector, its magnitude is independent of its direction. Algebraically, a vector is written as a row or column of real numbers (or expressions). And each number is called an element of the vector and represents a distinct dimension (coordinate) of the vector and the vector space where the vector lies. The total number of the elements of a vector is called the order or size of the vector. Thus, if a vector has n elements, it is a vector of order n or size n or just an n-element vector. Since a vector has magnitude and direction, geometrically, it is shown as an arrow in a vector space. For convenience, a vector is often written as a column vector in vector theory, unless otherwise indicated.

© The Author(s), under exclusive license to Springer Nature Switzerland AG 2025

B. T. Ro, *The Geometry of Accounting*, Springer Texts in Business and Economics,
https://doi.org/10.1007/978-3-031-88892-2_6

A row vector (**v**) of order n is written in general notations as

$$\mathbf{v} = (v_1, v_2, \ldots, v_n)$$

where v_n is the nth element of **v**. And the magnitude of **v** is defined and computed as follows:

$$|\mathbf{v}| = \left(v_1^{\,2} + v_2^{\,2} + \ldots + v_n^{\,2}\right)^{1/2} = \left(\sum v_j^{\,2}\right)^{1/2}$$

As seen, the magnitude of **v** equals the square root of the sum of squares of its elements. If, for example, $\mathbf{v} = (3, 4)$, $|\mathbf{v}| = (3^2 + 4^2)^{1/2} = 5$.

There are different kinds of vectors. If the elements of a vector are arranged into a single row (column), the vector is called a row (column) vector. For convenience, a vector is often written as a column vector, unless otherwise indicated. If all elements of a vector are zero, the vector is called a null vector, which represents the point of origin of the vector space. If a vector starts at this point, it is called a position vector; otherwise, it is a free vector. A free vector can be made a position vector, or vice versa, by displacing it in parallel without changing its magnitude and direction. It is a convention in vector theory that vectors representing points, lines, and spaces are all considered position vectors (unless indicated otherwise). Two identical vectors can start at different points or lie at different locations in the same vector space since vectors can be displaced in parallel. If a vector lies on a coordinate of the vector space, it is called a coordinate vector. If a vector points in the positive (negative) direction of a reference point or a reference vector, it is called a positive (negative) vector.

As with real numbers, vectors can be added or subtracted if they are conformable for addition or subtraction, that is, if they have the same order. Vector addition yields a sum vector, and vector subtraction yields a difference vector. Vectors can also be multiplied if they are conformable for multiplication, that is, if they have the same number of elements. Unlike vector addition or subtraction, however, the result of vector multiplication is a scalar number, not a vector. While vectors can be multiplied, a vector cannot be divided by another vector because vectors have a direction and dividing one direction by another makes little sense. Yet the magnitude of a vector can be divided by the magnitude of another vector. If, for example, **u** and **v** are two different vectors, the division **u** / **v** does not yield a meaningful result but |**u**| / |**v**| does.

A Euclidean vector space with Cartesian coordinates has an origin—the point at which the coordinates intersect. In this space, a vector is an *ordered* array, not just an array, of real numbers. Each of these numbers has a geometric meaning. A point is a directed location from the point of origin and, hence, is a vector. In an n-dimensional vector space with n coordinates, a point is an ordered array of n real numbers. A line is the shortest distance between two points. Thus, it is defined by a pair of vectors, one for each endpoint of the line. If a line starts at the point of origin, it is often written as a single vector, rather than a pair of vectors, since the point of origin is a null vector. A space is defined by three noncollinear points, or a minimum of one line and one point. Each point is a vector so that a vector space is defined by a minimum of three noncollinear vectors.

Conceptually, a vector space is considered a collection of all possible vectors that can be generated by a pair of binary vector operations—scalar multiplication and addition, which is called a linear combination of vectors. This is a mathematical operation of scaling multiple vectors with a scalar (or a set of scalars) and adding up the resultant scaled vectors. The result is a new vector. As such, a vector in a vector space is considered an element of the vector space. Every vector can be generated as a linear combination of some other vectors in the same vector space. The vectors used to perform such a linear combination are called basis vectors. A vector space is a geometric space generated as various linear combinations of such basis vectors. In other words, basis vectors as a set can generate another vector and span a vector space.

The Vectors and Vector Spaces for Accounting Events

Accounting events (transactions) have three intrinsic elements—debit, credit, and size (monetary volume), denoted as d, c, and q in monetary quantity (see Chaps. 1 and 2). Each of the elements is shown by a real number and represents a distinct dimension of accounting events. Thus, an accounting event can be written as a vector: (d, c) or (d, c, q), depending on the dimensionality of the vector space where the accounting event is presented. This vector is a position vector as it starts at the origin (zero monetary quantity) of the vector space. Each element of the accounting event is considered a point on a coordinate of the vector space and is hence a vector—a coordinate vector, to be exact. This vector is also a position vector and starts at the point of origin. These coordinate vectors are different from the vector (d, c) or (d, c, q), which represents the accounting event itself. This two- or three-element vector is also a position vector but is not a coordinate vector.

Characterizing accounting events and their debits and credits as vectors makes it possible to present accounting events in a multidimensional vector space. This, in turn, makes it possible to examine the features and properties of accounting events and explore the features of double entry accounting based on vectors and theories thereof.

The Vectors of Accounting Events in Two-Dimensional Space Accounting events can be presented in a two-dimensional vector space with two coordinates, one for the debit of the events and the other for the credit. Since there is no third dimension in the two-dimensional vector space, the size dimension of accounting events is suppressed. The point at which the debit and credit coordinates intersect is the point of origin—the point of zero monetary quantity. In the two-dimensional debit-credit vector space, all vectors of accounting events are of order 2 or two-element vectors. The vector of the debit and credit elements d and c respectively of accounting events represents a point or line on the respective debit and credit coordinates. They are coordinate vectors, written as $\mathbf{d} = (\mathrm{d}, 0)$ and $\mathbf{c} = (0, \mathrm{c})$. Hereafter, the former is referred to as the debit coordinate vector, or just the debit vector, of accounting events and the latter as the credit coordinate vector, or just the credit vector. They both are position vectors. As the debit and credit coordinates are orthogonal, so are the debit and credit vectors.

Every accounting event has a debit vector and a credit vector. The dual vectors have the same magnitude and act in perpendicular directions under the double entry principle. Together, they generate another vector, which represents the accounting event itself. This vector is called the double entry (DE) vector of accounting events, denoted as **D** hereafter. This vector is generated by **d** and **c** as a pair of basis vectors according to the double entry principle. Unlike the debit and credit vectors, **D** is not a coordinate vector. It lies *off* the debit and credit coordinates.

Given the above, every accounting event has the following triple position vectors in the two-dimensional debit-credit vector space:

$$\mathbf{d} = (\mathrm{d}, 0)$$

$$\mathbf{c} = (0, \mathrm{c})$$

$$\mathbf{D} = \mathbf{d} + \mathbf{c} = (\mathrm{d}, \mathrm{c}) \quad (|\mathbf{d}| = |\mathbf{c}|)$$

The DE vector **D** is a linear combination of **d** and **c**, each scaled by a factor of 1. As a pair of basis vectors, **d** and **c** generate **D**. And **d** and **c** are the component vectors of **D**. As seen, **D** is a sum vector. It is equivalent to the double entry (DE) point in Sect. 3.4, Chap. 3, and the DE line there. **D** represents the point (d, c) in the two-dimensional (debit-credit) vector space or the line between the two points (0,0 and (d,c). The point (d,c) lies off the debit and credit coordinates. Likewise, **D** also lies off the coordinates. Thus, **D** is not a coordinate vector.

The coordinate vectors **d** and **c** define a square—a DE square—in the two-dimensional vector space. And **D** is equal to the diagonal of the square and divides the square into two contrasting yet symmetric triangles. Thus, the magnitude of **D** is computed according to the Pythagorean theorem or the algebraic definition of the magnitude of vectors in general seen before, as follows:

$$|\mathbf{D}| = \left(\mathrm{d}^2 + \mathrm{c}^2\right)^{1/2}$$

For a given accounting event, the elements d and c of **D** can be positive or negative numbers, depending on whether the event increases or decreases the debit and credit accounts. Regardless, for every accounting event, $|\mathbf{d}| = |\mathbf{c}|$ or $|\mathrm{d}| = |\mathrm{c}|$ under the double entry principle. That is, **d** and **c** have the same magnitude and, as coordinate vectors, they are mutually orthogonal. Thus, **D** has an angle of 45° with both **d** and **c** and has a slope of 1 (tangent 45° = 1). This means that the relation between **d** and **c** is a perfectly balanced one and **D** indicates this relationship. It is equivalent to the DE line in Chap. 3. Conceptually, **D** is considered to reflect the duality and balance features of accounting events and the contrasting yet symmetric nature of double entries. Or, more generally, it reflects the duality and balance features of double entry accounting.

As a pair of basis vectors, **d** and **c** not only generate **D** but also span a vector space. This space is a square and, hereafter, referred to as the double entry (DE) vector space of accounting events. It is equivalent to the DE square in Chap. 3. As **D** is equal to the diagonal of the DE square, it is a diagonal vector.[1] As a square, the DE vector space reflects the duality and balance features of accounting events and the contrasting yet symmetric nature of double entries, just as the DE vector **D** does.

The Vectors of Accounting Events in Three-Dimensional Space Accounting events can be presented in a three-dimensional vector space with three coordinates. In addition to the debit and credit dimensions, there is another dimension, which represents the (monetary) size of accounting events. The size element q of accounting events is a point or line, thus a vector, on the size coordinate. The debit and credit elements of accounting events are a point or line, each being a coordinate vector, as they are in the two-dimensional vector space before. As a set of basis vectors, the three coordinate vectors of accounting events generate another vector—the double entry (DE) vector **D** of the events. Like the coordinate vectors, **D** is a position vector and starts at the point of origin.

Given the above, every accounting event has the four position vectors of order 3 in the three-dimensional vector space:

$$\mathbf{d} = (\mathrm{d}, 0, 0)$$

$$\mathbf{c} = (0, \mathrm{c}, 0)$$

$$\mathbf{q} = (0, 0, \mathrm{q})$$

$$\mathbf{D} = \mathbf{d} + \mathbf{c} + \mathbf{q} = (\mathrm{d}, \mathrm{c}, \mathrm{q}) \quad (|\mathrm{d}| = |\mathrm{c}| = \mathrm{q})$$

The DE vector **D** is a linear combination of **d**, **c**, and **q**, each scaled by a factor of 1. And **d**, **c**, and **q** are the component vectors of **D**. As coordinate vectors, **d**, **c**, and **q** are mutually orthogonal (perpendicular) and have the same magnitude $|\mathbf{d}| = |\mathbf{c}| = |\mathbf{q}|$ or $|\mathrm{d}| = |\mathrm{c}| = \mathrm{q}$ under the double entry principle. Thus, the magnitude of **D** is computed according to the Pythagorean theorem, or the algebraic definition of the magnitude of vectors indicated before, as follows:

$$|\mathbf{D}| = \left(\mathrm{d}^2 + \mathrm{c}^2 + \mathrm{q}^2\right)^{1/2}$$

The elements d and c of accounting events are negative numbers if they mean a decrease in the debit and credit accounts. Unlike these elements, however, the

[1] A diagonal vector is a vector that connects the two diagonal vertices of a polygon such as square, rectangle, parallelogram, or trapezoid in a multidimensional vector space.

element q is always positive ($q > 0$) since the monetary size (volume) of accounting events cannot be negative or zero. No accounting event can take place with a zero or negative monetary volume.

As a linear combination of **d**, **c**, and **q**, the DE vector **D** represents the relationship among the three component vectors, which is a perfectly balanced one. Conceptually, **D** reflects the duality and balance features of accounting events and the contrasting yet symmetric nature of double entries. It is equivalent to the DE vector in the two-dimensional space before and the DE line in Sect. 4.4, Chap. 4. For a given accounting event, all those vectors are interrelated, co-exist, and act together in perpendicular directions under the double entry principle.

As a set of basis vectors, **d**, **c**, and **q** generate not only **D** but also span a vector space in the three-dimensional vector space. This vector space is a cube since **d**, **c**, and **q** have the same magnitude and are orthogonal. It is the double entry (DE) vector space of accounting events. It is equivalent to the DE vector **D** before and the DE cube and DE line in Chap. 4.

6.2 The Settings of Vector Spaces for Accounting Events

This section discusses the general settings of two- and three-dimensional vector spaces where accounting events can be presented in vector form according to the double entry principle. The vectors of accounting events are all of order 2 or order 3, depending on the dimensionality of the vector space where the events are presented.

The Setting of the Two-Dimensional Vector Space for Double Entries

The two-dimensional vector space where accounting events are presented according to the double entry principle has a Cartesian space with two coordinates, one for the debit dimension of accounting events and the other for the credit dimension. The size dimension of accounting events is suppressed as there is no third dimension for it.

Figure 6.1, adapted from Fig. 3.1, gives an example of a two-dimensional vector space where accounting events can be presented. The point O is the point of origin (zero monetary quantity). With this point set, the two coordinates divide the entire debit-credit coordinate system into four quadrants (QI to QIV). The arrows shown are all position vectors of order 2 and start at the point of origin. The arrow on the positive side of the debit coordinate is the debit vector $\mathbf{d} = (d, 0)$, and the one on the negative side is the debit vector $\mathbf{b} = (b, 0)$. The former represents a debit account as an increase, and the latter a debit account as a decrease. The two arrows on the credit coordinate represent the corresponding credit vectors: $\mathbf{c} = (0, c)$ as an increase and

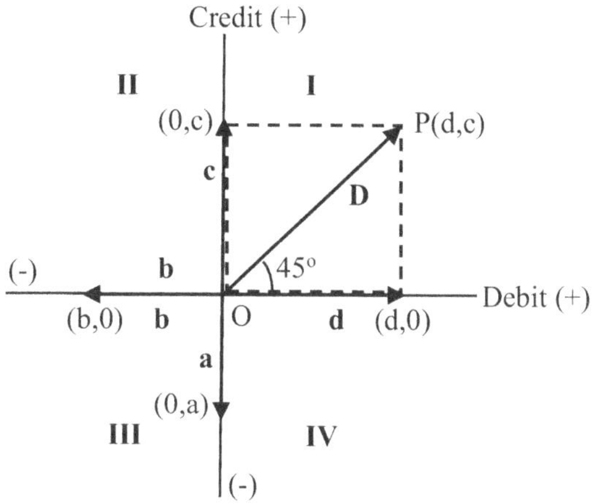

Fig. 6.1 The setting of a two-dimensional vector space for the presentation of accounting events

a = (0, a) as a decrease.[2] The arrow **D** is the DE vector of the accounting event. This event is assumed to be of type 1 and increase the debit and credit accounts alike. Thus, **D** lies in QI and points northeast. Since **d** and **c** each mean an increase in monetary quantity, they are positive coordinate vectors. **b** and **a** are the opposite—negative coordinate vectors as they mean a decrease in monetary quantity. This does not mean that the decrease is negative in substance since a monetary quantity cannot be intrinsically negative. As a negative vector, **b** points in the opposite direction of **d**, and **a** in the opposite direction of **c**. For an accounting event, **d** and **c** have the same magnitude (|**d**| = |**c**|) and point in perpendicular directions under the double entry principle. As a pair of basis vectors, they generate **D**, which has an angle of 45° with the debit and credit coordinates and has a slope of 1 as tangent 45° = 1. As such, **D** reflects the perfectly balanced relationship between the debit and credit of the accounting event. More generally, it reflects the duality and balance features of the accounting event and the contrasting yet symmetric nature of double entries. The dashed-line square in the figure is the DE vector space (DE square) of the accounting event. It is defined by **d** and **c** as a pair of basis vectors. It is a square since **d** and **c** are orthogonal and |**d**| = |**c**| under the double entry principle. It also reflects the same duality and balance features of the accounting event as **D** does.

The accounting event in the figure is assumed to be of type 1 and increase the debit and credit accounts alike. Thus, the DE vector space and DE vector **D** lie in QI. If an accounting event is of the same type but decreases the account on each side, the DE vector space and DE vector **D** lie in QIII. And **D** points southwest. If neither is the case, they lie in QII or QIV.

In general, the form of the DE vector of type 1 accounting events and its location in the two-dimensional vector space depend on whether the accounting events

[2]The coordinate vector notations **d**, **c**, **b**, and **a** are consistent with the notations of the cell blocks of the tableau of the double entry system in Figure 2.3. As such, they mean what the notations of the cell blocks there mean.

increase or decrease the debit and credit accounts, respectively. Shown below are the possible forms of **D** for such accounting events and the possible locations of the DE vector.

$$\begin{aligned}\mathbf{D} &= \mathbf{d} + \mathbf{c} = (\mathrm{d}, \mathrm{c}) \quad (\mathrm{d} = \mathrm{c}, \mathbf{D} \text{ in QI}) \\ &= \mathbf{d} + \mathbf{a} = (\mathrm{d}, \mathrm{a}) \quad (\mathrm{d} = |\mathrm{a}|, \mathbf{D} \text{ in QIV}) \\ &= \mathbf{b} + \mathbf{c} = (\mathrm{b}, \mathrm{c}) \quad (|\mathrm{b}| = \mathrm{c}, \mathrm{D} \text{ in QII})\end{aligned}$$

or

$$= \mathbf{b} + \mathbf{a} = (\mathrm{b}, \mathrm{a}) \quad (\mathrm{b} = \mathrm{a}, \mathbf{D} \text{in QIII})$$

where b and a are negative numbers and represent the debit and credit accounts respectively as a decrease. Thus, they are shown in absolute value. Regardless of its form, **D** is a linear combination of those coordinate vectors shown. Each pair of the adjacent coordinate vectors above, such as **d** and **c** or **d** and **a**, represents a type 1 accounting event.

For a given accounting event, all the vectors including **D** point in different directions, meaning that they are independent in a mathematical sense. From the perspective of double entry accounting, however, they are *not* because they are interrelated and mutually dependent. None of them can exist by itself. For a given accounting event, they always co-exist and act together under the double entry principle. And, if one of them is known or determined, the others are also.

If an accounting event is of type 2, 3, or 4, it has component transactions, which are executed in bundle as the event is executed. Hence, **d**, **c**, and **D** each are not one continuous vector but divided into multiple discrete vectors, one for each component transaction. If, for example, an accounting event has two component transactions and increases the debit and credit accounts alike, the vectors of this event are written as

$$\mathbf{d} = \mathbf{d}_1 + \mathbf{d}_2 = (\mathrm{d}_1, 0) + (\mathrm{d}_2, 0) = (\mathrm{d}_1 + \mathrm{d}_2, 0)$$

$$\mathbf{c} = \mathbf{c}_1 + \mathbf{c}_2 = (0, \mathrm{c}_1) + (0, \mathrm{c}_2) = (0, \mathrm{c}_1 + \mathrm{c}_2)$$

$$\mathbf{D} = \mathbf{d} + \mathbf{c} = (\mathrm{d}_1 + \mathrm{d}_2, \mathrm{c}_1 + \mathrm{c}_2)$$

or

$$\mathbf{D}_1 = \mathbf{d}_1 + \mathbf{c}_1 = (\mathrm{d}_1, \mathrm{c}_1)$$

$$\mathbf{D}_2 = \mathbf{d}_2 + \mathbf{c}_2 = (\mathrm{d}_2, \mathrm{c}_2)$$

$$\mathbf{D} = \mathbf{D}_1 + \mathbf{D}_2 = (\mathrm{d}_1 + \mathrm{d}_2, \mathrm{c}_1 + \mathrm{c}_2)$$

For type 2, 3, or 4 accounting events, the multiple accounts on the debit or credit side (or both sides) of the events may or may not change in the same direction. If they

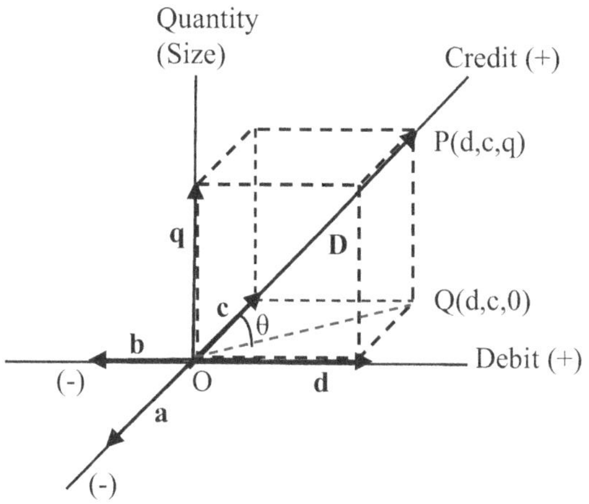

Fig. 6.2 The setting of a three-dimensional vector space for the presentation of accounting events

do not, the discrete component vectors of **d** and **c** respectively do not point in the same direction. Nor do they lie in the same quadrant of the two-dimensional vector space. Therefore, **d** and **c** cannot be linearly added to get **D** as a single DE vector. Such accounting events are represented by a set of discrete DE vectors, one for each component transaction. Thus, **D** is divided into multiple discrete vectors.

The Setting of the Three-Dimensional Vector Space for Double Entries

In the three-dimensional (debit-credit-size) vector space, there exists a third coordinate and all vectors are of order 3. Figure 6.2, adapted from Fig. 4.1, gives an example of a three-dimensional vector space where accounting events can be presented. The coordinate *Quantity (Size)* represents the size dimension of accounting events. It has no negative side since the monetary size q of accounting events cannot be negative or even zero ($q > 0$). No accounting event can have a zero or negative monetary volume.

As seen, there are five coordinate vectors, all of which are of order 3: **d** = (d, 0, 0), **c** = (0, c, 0), **b** = (b, 0, 0), **a** = (0, a, 0), and **q** = (0, 0, q). The size coordinate vector **q** is always a positive vector for the reason indicated above. The dashed-line cube is the DE vector space of an accounting event, and the diagonal vector **D** (arrow OP) is the DE vector of the event. The coordinate vectors **d**, **c**, and **q** have the same magnitude and point in perpendicular directions under the double entry principle. Although mutually perpendicular, they are interrelated and not independent under the double entry principle. All those vectors are position vectors and start at the point O (zero monetary quantity). As a set, the adjacent triple coordinate vectors (e.g., **d**, **c**, and **q**) can be basis vectors. Together, they generate another vector such as **D** and span a vector space such as the dashed-line DE vector space. Since the three coordinate vectors have the same magnitude and point in perpendicular directions,

D has an angle of 35.2645° ($\theta \approx 35.2645°$) with the bottom face of the DE vector space and has a slope of 0.7071, which is equal to tangent 35.2645°. **D** is equivalent to the DE line of accounting events in Sect. 4.4, Chap. 4, and the diagonal vertex P(d,c,q) is equivalent to the DE point of accounting events there.

The accounting event in the figure is assumed to be of type 1 and increase the debit and credit accounts alike. Thus, both the DE cube and the DE vector lie in the first octant (OI) of the three-dimensional vector space. In general, the DE vector **D** and DE cube of type 1 accounting events can lie in any octant, depending on whether the events increase or decrease the debit and credit accounts, respectively. **D** for such accounting events and its location can vary as shown below.

$$\begin{aligned}\mathbf{D} &= \mathbf{d} + \mathbf{c} + \mathbf{q} = (\mathrm{d}, \mathrm{c}, \mathrm{q}) \quad (\mathrm{d} = \mathrm{c} = \mathrm{q} \text{ in OI}) \\ &= \mathbf{d} + \mathbf{a} + \mathbf{q} = (\mathrm{d}, \mathrm{a}, \mathrm{q}) \quad (\mathrm{d} = |\mathrm{a}| = \mathrm{q} \text{ in OIV}) \\ &= \mathbf{b} + \mathbf{c} + \mathbf{q} = (\mathrm{b}, \mathrm{c}, \mathrm{q}) \quad (|\mathrm{b}| = \mathrm{c} = \mathrm{q} \text{ in OII})\end{aligned}$$

or

$$= \mathbf{b} + \mathbf{a} + \mathbf{q} = (\mathrm{b}, \mathrm{a}, \mathrm{q}) \quad (|\mathrm{b}| = |\mathrm{a}| = \mathrm{q} \text{ in OIII})$$

If an accounting event is of type 2, 3, or 4 and has multiple accounts on the debit or credit side (or both sides), it has component transactions. Thus, each coordinate vector of the accounting event is divided into a set of discrete vectors. So is **D**. If, for example, an accounting event has two component transactions and increases the accounts on both debit and credit sides alike, the vectors of the accounting event are written as

$$\mathbf{d} = \mathbf{d}_1 + \mathbf{d}_2 = (\mathrm{d}_1, 0, 0) + (\mathrm{d}_2, 0, 0) = (\mathrm{d}_1 + \mathrm{d}_2, 0, 0)$$

$$\mathbf{c} = \mathbf{c}_1 + \mathbf{c}_2 = (0, \mathrm{c}_1, 0) + (0, \mathrm{c}_2, 0) = (0, \mathrm{c}_1 + \mathrm{c}_2, 0)$$

$$\mathbf{q} = \mathbf{q}_1 + \mathbf{q}_2 = (0, 0, \mathrm{q}_1) + (0, 0, \mathrm{q}_2) = (0, 0, \mathrm{q}_1 + \mathrm{q}_2)$$

$$\mathbf{D} = \mathbf{d} + \mathbf{c} + \mathbf{q} = (\mathrm{d}_1 + \mathrm{d}_2, \mathrm{c}_1 + \mathrm{c}_2, \mathrm{q}_1 + \mathrm{q}_2)$$

or

$$\mathbf{D}_1 = \mathbf{d}_1 + \mathbf{c}_1 + \mathbf{q}_1 = (\mathrm{d}_1, \mathrm{c}_1, \mathrm{q}_1)$$

$$\mathbf{D}_2 = \mathbf{d}_2 + \mathbf{c}_2 + \mathbf{q}_2 = (\mathrm{d}_2, \mathrm{c}_2, \mathrm{q}_2)$$

$$\mathbf{D} = \mathbf{D}_1 + \mathbf{D}_2$$

The vectors **d**, **c**, and **q** lie on the positive side of the respective coordinates, and **D** lies in OI. While all those vectors point in different directions, they are not independent since they are interrelated and act together according to the double entry principle. For a given accounting event, they co-exist and co-vary and are hence not independent.

6.3 Accounting Events in Two-Dimensional Vector Space

The presentation of accounting events in the two-dimensional (debit-credit) vector space can be explained with specific transaction examples. This section explains it with four transaction examples in Chap. 3. Since there are only two dimensions in the two-dimensional space, the size dimension of accounting events is suppressed.

Type 1 Events: A Single Account (Entry) on Both Sides

Consider the type 1 accounting event (in Chap. 3)—that a firm sells goods for $100 in cash. This event has a single account (entry) on both debit and credit sides and increases the account on each side. It has no component transactions. Figure 6.3 presents the event in the two-dimensional (debit-credit) vector space. The arrows **d** and **c** are the debit and credit vectors of the accounting event, and **D** is the DE vector. They are all position vectors of order 2 and start at the point O as their common origin. As the accounting event increases the debit and credit accounts alike, both **d** and **c** lie on the positive side of the debit and credit coordinates, respectively. And **D** lies in QI of the two-dimensional vector space. The three vectors are written as

$$\begin{aligned} \mathbf{d} &= (100, 0) \\ \mathbf{c} &= (0, 100) \\ \mathbf{D} &= \mathbf{d} + \mathbf{c} = (100, 100) \end{aligned}$$

As coordinate vectors, **d** and **c** are orthogonal ($\theta = 90°$) and $|\mathbf{d}| = |\mathbf{c}|$ under the double entry principle. As a pair of basis vectors, they span the shaded DE vector space, which is a square. This vector space is equivalent to the DE square of the same accounting event in Sect. 3.2, Chap. 3. As a pair of basis vectors, **d** and **c** also

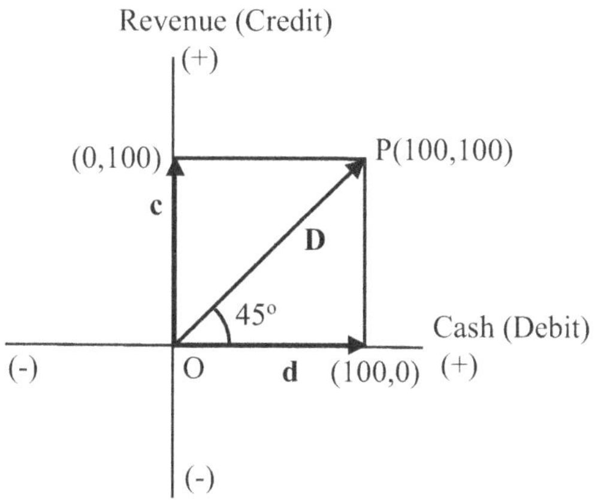

Fig. 6.3 Presenting a type 1 accounting in the two-dimensional vector space

generate **D**. This DE vector is equal to the diagonal of the DE vector space. It is equivalent to the DE line of the accounting event in Sect. 3.4, Chap. 3. As a diagonal vector, it has a slope of 1 as it has an angle of 45° with **d** and **c**. It reflects the duality and balance features of the accounting event.

The size of the accounting event is equal to $100 in monetary terms. If measured geometrically in the two-dimensional vector space, it is equal to the magnitude (length) of the DE vector or the size (Q_S) of the DE vector space. Each is computed below.

$$|\mathbf{D}| = \left(|\mathbf{d}|^2 + |\mathbf{c}|^2\right)^{1/2} = 100\sqrt{2} = 141.4214$$

and

$$Q_S = |\mathbf{d}||\mathbf{c}| = \mathrm{d.c} = 100.100 = 100^2\mathrm{au}^2$$

The magnitude of **D** can be computed based on the algebraic definition of the magnitude of vectors in general (seen in Sect. 6.1) or according to the Pythagorean theorem. It is equal in magnitude to the line-based measure (Q_L) of the size of the accounting event in Sect. 3.4, Chap. 3. And Q_S is equal to the space-based measure of the size of the event in Sect. 3.2, Chap. 3. As a diagonal vector, **D** divides the DE square into two symmetric right triangles of equal size. Thus, |**D**| can also be computed according to the law of cosine or the law of sine respectively as follows:

$$|\mathbf{D}| = |\mathbf{d}|/\text{cosine } 45^{\circ} = 141.4214$$

or

$$|\mathbf{D}| = |\mathbf{c}|/\text{sine } 45^{\circ} = 141.4214$$

The two geometric measures |**D**| of the size of the accounting event are the same since cosine 45° = sine 45°. Hence, they can be converted to the same monetary size of the event as follows:

$$\begin{aligned} \mathrm{q} &= |\mathbf{d}| = \left(|\mathbf{D}|^2 - |\mathbf{c}|^2\right)^{1/2} = \left(141.42136^2 - 100^2\right)^{1/2} \\ &= 100 \end{aligned}$$

or

$$\begin{aligned} \mathrm{q} &= |\mathbf{D}|\text{sine } 45^{\circ} \text{ or } |\mathbf{D}| \text{ cosine } 45^{\circ} \\ &= Q_L\text{sine } 45^{\circ} \text{ or } Q_L\text{cosine } 45^{\circ} \\ &= (141.42136)(0.70710678) = 100. \end{aligned}$$

Type 2 Events: Multiple Accounts (Entries) on the Debit Side

Somewhat more complex than the previous type 1 transaction is the type 2 transaction example—that a firm sells goods for $100 by receiving $60 in cash and holding $40 on account for future collection. This transaction has two debit accounts (cash and account receivable) and increases both of them. Thus, it has two component transactions, one for the cash part and the other for the account-receivable part. Figure 6.4 presents the accounting event in the two-dimensional vector space. Figure 6.4a shows the DE vector space (large shaded square) of the event along with the DE vector spaces (small squares) of the two component transactions. The small square in the lower-left corner is the DE vector space of the cash part of the accounting event, and the small square in the upper-right corner is the DE vector space of the account-receivable part. The two component DE vector spaces are connected at the joint vertex (60,60). As the event increases the debit and credit accounts alike, the DE vector space lies in QI. Figure 6.4b shows the DE vector space of each component transaction, by displacing $\mathbf{d}_2$ and $\mathbf{c}_2$ as position vectors starting at the point of origin. Displacing them in parallel is possible since vectors in general can be displaced in parallel without changing their magnitude and direction.

As the accounting event has two component transactions, **d** and **c** each are divided into two discrete vectors: $\mathbf{d}_1 = (60, 0)$ and $\mathbf{d}_2 = (40, 0)$, and $\mathbf{c}_1 = (0, 60)$ and

Fig. 6.4 Presenting a type 2 accounting event in the two-dimensional vector space

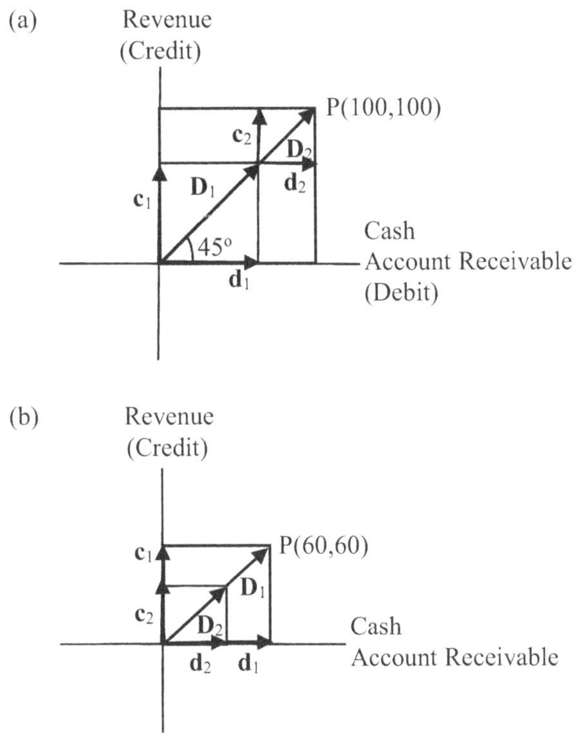

$\mathbf{c}_2 = (0, 40)$. The vectors $\mathbf{d}_2$ and $\mathbf{c}_2$ in Fig. 6.4a may be written as $\mathbf{d}_2 = (100–60, 0)$ and $\mathbf{c}_2 = (0, 100–60)$. If they are displaced in parallel as indicated, they become position vectors starting at the point of origin and can thus be written as $\mathbf{d}_2 = (40, 0)$ and $\mathbf{c}_2 = (0, 40)$, which is what is Fig. 6.4b shows.[3] As a pair, $\mathbf{d}_1$ and $\mathbf{c}_1$ represent the cash part of the accounting event, and $\mathbf{d}_2$ and $\mathbf{c}_2$ represent the account-receivable part. Given this, the DE vector $\mathbf{D}$ of the accounting event is also divided accordingly, $\mathbf{D}_1$ and $\mathbf{D}_2$. Shown below are all these vectors.

$$\begin{aligned}
\mathbf{d} &= \mathbf{d}_1 + \mathbf{d}_2 = (60 + 40, 0) = (100, 0)\\
\mathbf{c} &= \mathbf{c}_1 + \mathbf{c}_2 = (0, 60 + 40) = (0, 100)\\
\mathbf{D}_1 &= \mathbf{d}_1 + \mathbf{c}_1 = (60, 60)\\
\mathbf{D}_2 &= \mathbf{d}_2 + \mathbf{c}_2 = (40, 40)\\
\mathbf{D} &= \mathbf{D}_1 + \mathbf{D}_2 = (60, 60) + (40, 40) = (100, 100)\\
&= \mathbf{d} + \mathbf{c} = (100, 100)
\end{aligned}$$

Since the accounting event increases the debit and credit accounts alike, $\mathbf{D}_1$ and $\mathbf{D}_2$ are collinear and lie in QI. Thus, they can be linearly added to get $\mathbf{D}$ as a single vector. As a pair of basis vectors, $\mathbf{d}$ and $\mathbf{c}$ generate $\mathbf{D}$ and span the DE square in Fig. 6.4a. Likewise, $\mathbf{d}_1$ and $\mathbf{c}_1$ ($\mathbf{d}_2$ and $\mathbf{c}_2$) can also be such basis vectors, generate $\mathbf{D}_1$ ($\mathbf{D}_2$), and span the respective component DE square in Fig. 6.4b.

The size (quantity) of the accounting event in monetary terms is \$100, which is equal to the magnitude (length) of $\mathbf{d}$ or $\mathbf{c}$. If measured geometrically in the vector space, it is equal to the magnitude of $\mathbf{D}_1$ and $\mathbf{D}_2$ combined. Each of these DE vectors divides the related component DE square into two symmetric right triangles of equal size as shown in the figure. Thus, the size of the accounting event as measured by the magnitude of $\mathbf{D}$ is computed as follows:

$$\begin{aligned}
|\mathbf{D}_1| &= \left(|\mathbf{d}_1|^2 + |\mathbf{c}_1|^2\right)^{1/2} = 60\sqrt{2} = 84.8528\\
|\mathbf{D}_2| &= \left(|\mathbf{d}_2|^2 + |\mathbf{c}_2|^2\right)^{1/2} = 40\sqrt{2} = 56.5685\\
|\mathbf{D}| &= |\mathbf{D}_1| + |\mathbf{D}_2| = 141.4214
\end{aligned}$$

or

$$\begin{aligned}
|\mathbf{D}| &= (|\mathbf{d}_1| + |\mathbf{d}_2|)/\text{cosine } 45^\circ \text{ or } (|\mathbf{c}_1| + |\mathbf{c}_2|)/\text{sine } 45^\circ\\
&= (60 + 40)/0.70710678 = 141.4214
\end{aligned}$$

Alternatively, the size of the accounting event can also be measured by the size of the DE vector space as

[3] Hereafter, all such component or discrete vectors are written and shown in the same way (unless otherwise indicated).

$$\begin{aligned} Q_S &= (|\mathbf{d}_1| + |\mathbf{d}_2|)(|\mathbf{c}_1| + |\mathbf{c}_2|) = (60 + 40)(60 + 40) \\ &= 100^2 \mathrm{au}^2 \end{aligned}$$

Although different in magnitude, the line- and space-based measures of the size of the accounting event are equivalent as they both are a geometric measure of the size of the same accounting event. They are also equivalent to the space-based measure Q_S of the size of the event in Sect. 3.2, Chap. 3.

As the line- and space-based measures of size are equivalent, they can be converted to the same monetary size of the accounting event:

$$\begin{aligned} q &= |\mathbf{d}| = \left[|\mathbf{D}|^2 - (|\mathbf{c}_1| + |\mathbf{c}_2|)^2\right]^{1/2} \\ &= \left(141.4214^2 - 100^2\right)^{1/2} = 100 \end{aligned}$$

or

$$\begin{aligned} q &= |\mathbf{d}_1| + |\mathbf{d}_2| = |\mathbf{D}|\text{cosine } 45^{\mathrm{o}} \text{ or } |\mathbf{D}| \text{ sine } 45^{\mathrm{o}} \\ &= (141.4214)(0.70710678) = 100 \end{aligned}$$

or

$$q = (Q_S)^{1/2} = 100$$

Type 3 Events: Multiple Accounts (Entries) on the Credit Side

To illustrate how to present this type of accounting events in the two-dimensional vector space, refer back to the type 3 transaction example—that a firm purchases goods for \$100 by paying \$60 in cash and holding \$40 on account for future payment. This accounting event has one debit account (inventory asset) as an increase and two credit accounts, account payable as an increase and cash as a decrease. Thus, it has two component transactions, one for the cash part and the other for the account-payable part.

Figure 6.5 presents the accounting event in the two-dimensional vector space. The unshaded square is the DE vector space of the event. It lies across QI and QIV since the event decreases one of the credit accounts and increases the other. The small shaded square in QI represents the account-payable part of the event, and the one in QIV represents the cash part. Since the two credit accounts do not change in the same direction, the credit vectors of the accounting event are written as $\mathbf{c} = (0, 40)$ and $\mathbf{a} = (0, -60)$. The corresponding debit vectors are $\mathbf{d}_1 = (60, 0)$ and $\mathbf{d}_2 = (40, 0)$.[4] As a pair, $\mathbf{d}_1$ and $\mathbf{a}$ represent the cash part of the accounting event

[4] $\mathbf{d}_2$ may be written as $\mathbf{d}_2 = (100–60, 0)$. As indicated in the previous footnote, it is written as $\mathbf{d}_2 = (40, 0)$. Again, this is the case for all such vectors hereafter.

Fig. 6.5 Presenting a type 3 accounting event in the two-dimensional vector space

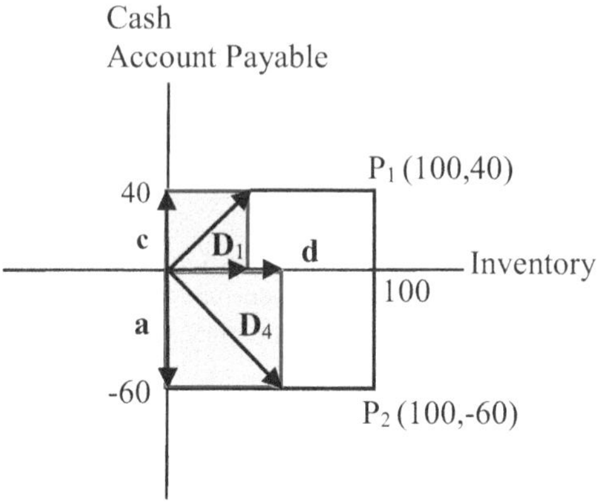

and, together, define the shaded DE square in QIV. And $\mathbf{d}_2$ and $\mathbf{c}$ represent the account-payable part and define the shaded DE square in QI. Given the two component transactions, the accounting event is represented by the pair of DE vectors: $\mathbf{D}_1 = (40, 40)$ and $\mathbf{D}_4 = (60, -60)$, where the subscripts 1 and 4 denote the quadrants of the two-dimensional vector space in which the two vectors lie. As a pair of basis vectors, $\mathbf{d}_1$ and $\mathbf{a}$ ($\mathbf{d}_2$ and $\mathbf{c}$) generate $\mathbf{D}_4$ ($\mathbf{D}_1$) and span the shaded DE square in QIV (QI).

Shown below are all those vectors of the accounting event. Since $\mathbf{D}_1$ and $\mathbf{D}_4$ are not collinear, they cannot be linearly added to get $\mathbf{D}$ as a single vector.

$$\begin{aligned}
\mathbf{d} &= \mathbf{d}_1 + \mathbf{d}_2 = (60, 0) + (40, 0) = (100, 0) \\
\mathbf{c} &= (0, 40) \\
\mathbf{a} &= (0, -60) \\
\mathbf{D}_1 &= \mathbf{d}_2 + \mathbf{c} = (\mathrm{d}_2, \mathrm{c}) = (40, 40) \\
\mathbf{D}_4 &= \mathbf{d}_1 + \mathbf{a} = (\mathrm{d}_1, \mathrm{a}) = (60, -60)
\end{aligned}$$

If measured in monetary terms, the size of the accounting event is \$100, equal to $|\mathbf{d}_1| + |\mathbf{d}_2|$ or $|\mathbf{c}| + |\mathbf{a}|$. If measured geometrically in the vector space, it is equal to the magnitude of $\mathbf{D}$, which is the same as the magnitude of $\mathbf{D}_1$ and $\mathbf{D}_4$ combined. The magnitude of these vectors is computed below.

$$\begin{aligned}
|\mathbf{D}_1| &= \left(|\mathbf{d}_2|^2 + |\mathbf{c}|^2\right)^{1/2} = \left(40^2 + 40^2\right)^{1/2} = 56.5685 \\
|\mathbf{D}_4| &= \left(|\mathbf{d}_1|^2 + |\mathbf{a}|^2\right)^{1/2} = \left(60^2 + |-60|^2\right)^{1/2} = 84.8528 \\
|\mathbf{D}| &= |\mathbf{D}_1| + |\mathbf{D}_4| = 40\sqrt{2} + 60\sqrt{2} = 141.4214
\end{aligned}$$

While $\mathbf{D}_1$ and $\mathbf{D}_4$ cannot be linearly added to get $\mathbf{D}$ as a single vector, their magnitudes can be since the magnitude of vectors is a directionless scalar quantity. The magnitude $|\mathbf{D}|$ equals the diagonal of the large unshaded DE square of the accounting event. It is equivalent to the line-based measure of the size of the accounting event in Sect. 3.4, Chap. 3.

The component vectors $\mathbf{D}_1$ and $\mathbf{D}_4$ each divide the related small DE square into two right triangles of equal size. Given the triangles, the magnitude $|\mathbf{D}|$ can also be computed as follows:

$$\begin{aligned} |\mathbf{D}| &= |\mathbf{D}_1| + |\mathbf{D}_4| = (40/\text{cosine } 45^\circ) + (60/\text{cosine } 45^\circ) \\ &= (40 + 60)/\text{cosine } 45^\circ \\ &= 141.4214 \end{aligned}$$

And the space-based measure of the size of the accounting event is computed as

$$\begin{aligned} Q_S &= (|\mathbf{d}_1| + |\mathbf{d}_2|)(|\mathbf{c}| + |\mathbf{a}|) = (60 + 40)(40 + |-60|) \\ &= 100^2 au^2 \end{aligned}$$

which is equal to the space-based measure Q_S of the size of the accounting event in Sect. 3.2, Chap. 3.

The alternative geometric measures above of the size of the accounting event are equivalent. Hence, they can be converted to the same monetary size of the event as follows:

$$\begin{aligned} q &= |\mathbf{d}| = \left(|\mathbf{D}|^2 - (|\mathbf{d}_1| + |\mathbf{d}_2|)^2\right)^{1/2} \\ &= \left(141.4214^2 - 100^2\right)^{1/2} = 100 \end{aligned}$$

or

$$\begin{aligned} q &= |\mathbf{d}_1| + |\mathbf{d}_2| = |\mathbf{D}|\text{cosine } 45^\circ \text{ or } |\mathbf{D}|\text{sine } 45^\circ \\ &= (141.4214)(0.70710678) = 100 \end{aligned}$$

or

$$q = (Q_S)^{1/2} = 100$$

Type 4 Events: Multiple Accounts (Entries) on Both Sides

More complex than the previous accounting events is the type 4 transaction example—that a firm recognizes an interest expense of \$330 on a bond liability, pays the cash of \$350 for the accrued interest on the bond liability, and amortizes a bond discount of \$100. This accounting event has three component transactions and two accounts on the debit and credit sides respectively, one as an increase and the other as a decrease. Figure 6.6a presents the accounting event in the two-dimensional

Fig. 6.6 Presenting a type 4 accounting event in the two-dimensional vector space

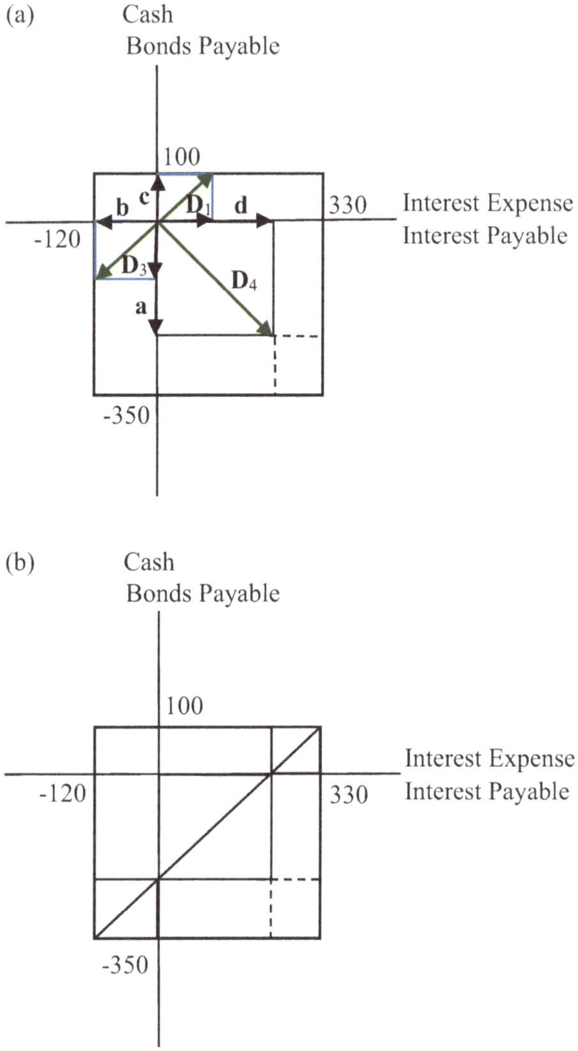

vector space. The large (unshaded) DE square is the DE vector space of the accounting event, and each small shaded square is the DE vector space of a component transaction. $\mathbf{D}_1$, $\mathbf{D}_3$, and $\mathbf{D}_4$ are the DE vectors of the respective component transactions, and the subscripts 1, 3, and 4 denote the quadrant of the vector space where those vectors lie. All the vectors shown are position vectors of order 2. And **d**, **c**, **b**, and **a** are all coordinate vectors, which are also of order 2. As the accounting event has three component transactions, it is represented by $\mathbf{D}_1$, $\mathbf{D}_3$, and $\mathbf{D}_4$ as a set. Since these DE vectors are not collinear, they cannot be linearly added to get **D** as a single DE vector. And **d** is divided into $\mathbf{d}_1$ and $\mathbf{d}_2$, and **a** into $\mathbf{a}_1$ and $\mathbf{a}_2$. As a pair, $\mathbf{d}_1$ and **c** represent the increase in the bond liability due to the amortization of the bond discount \$100. Similarly, **b** and $\mathbf{a}_1$ represent the payment in cash of the unpaid interest \$120, and $\mathbf{d}_2$ and $\mathbf{a}_2$ represent the payment of the newly accrued current-period interest \$230.

Given the three component transactions, the debit and credit vectors of the accounting event, and the DE vectors of the component transactions, are written as

$$\begin{aligned}
\mathbf{d} &= \mathbf{d}_1 + \mathbf{d}_2 = (100, 0) + (230, 0) = (330, 0) \\
\mathbf{c} &= (0, 100) \\
\mathbf{b} &= (-120, 0) \\
\mathbf{a} &= \mathbf{a}_1 + \mathbf{a}_2 = (0, -120) + (0, -230) = (0, -350) \\
\mathbf{D}_1 &= \mathbf{d}_1 + \mathbf{c} = (100, 100) \\
\mathbf{D}_3 &= \mathbf{b} + \mathbf{a}_1 = (-120, -120) \\
\mathbf{D}_4 &= \mathbf{d}_2 + \mathbf{a}_2 = (230, -230)
\end{aligned}$$

Each pair of the adjacent coordinate vectors can be basis vectors, generate a DE vector, and span a DE vector space. For example, $\mathbf{d}_1$ and **c** can generate $\mathbf{D}_1$ and span the shaded small DE square in QI.

The size of the accounting event in monetary terms equals \$450, which is equal to the combined magnitude $|\mathbf{d}| + |\mathbf{b}|$ or $|\mathbf{c}| + |\mathbf{a}|$. If measured geometrically in the vector space, it is equal to the combined magnitude (length) of $\mathbf{D}_1$, $\mathbf{D}_3$, and $\mathbf{D}_4$. It is computed as

$$\begin{aligned}
|\mathbf{D}| &= |\mathbf{D}_1| + |\mathbf{D}_e| + |\mathbf{D}_4| \\
&= 100\sqrt{2} + |-120|\sqrt{2} + 230\sqrt{2} \\
&= 636.3961
\end{aligned}$$

Although $\mathbf{D}_1$, $\mathbf{D}_3$, and $\mathbf{D}_4$ cannot be linearly added to get **D** as a single vector, their magnitudes can be.

To see more clearly what is shown above, the figure can be modified by repositioning the small DE squares as shown in Fig. 6.6b. Here, it is clear that the combined magnitude of $\mathbf{D}_1$, $\mathbf{D}_3$, and $\mathbf{D}_4$ equals the diagonal of the large DE square of the accounting event, which is what the magnitude $|\mathbf{D}|$ above stands for. This magnitude can also be computed according to the Pythagorean theorem or the law of cosine as follows:

$$|\mathbf{D}| = \left(450^2 + 450^2\right)^{1/2} = 450\sqrt{2} = 636.3961$$

or

$$\begin{aligned}
|\mathbf{D}| &= |\mathbf{D}_1| + |\mathbf{D}_3| + |\mathbf{D}_4| \\
&= (100/\text{cosine } 45^\circ + |-120|/\text{cosine } 45^\circ + 230/\text{cosine } 45^\circ) \\
&= 636.3961
\end{aligned}$$

The magnitude |**D**| is equivalent to the line-based measure Q_L of the size of the accounting event in Sect. 3.4, Chap. 3. It can be converted to the monetary size ($450) of the event as shown below.

$$\begin{aligned} |\mathbf{d}| + |\mathbf{b}| &= (|\mathbf{D}_1| + |\mathbf{D}_3| + |\mathbf{D}_4|) \text{ cosine } 45^{\circ} \\ &= \left(100\sqrt{2} + |-120|\sqrt{2} + 230\sqrt{2}\right) \text{ cosine } 45^{\circ} = 450 \\ &= |\mathbf{c}| + |\mathbf{a}| \end{aligned}$$

Alternatively, the size of the accounting event can also be measured by the size of the DE vector space:

$$\begin{aligned} Q_S &= (|\mathbf{d}_1| + |\mathbf{d}_2| + |\mathbf{b}|)(|\mathbf{c}| + |\mathbf{a}_1| + |\mathbf{a}_2|) \\ &= 450.450 = 450^2 \text{au}^2 \end{aligned}$$

The space-based measure above of the size measure of the accounting event is equivalent to the space-based measure Q_S of the size of the event in Sect. 3.2, Chap. 3.

6.4 Accounting Events in Three-Dimensional Vector Space

Presenting accounting events in the three-dimensional (debit-credit-size) vector space is more involved than it is in the two-dimensional vector space before since there is a third dimension to deal with. Yet it is more consistent with the multidimensional nature of accounting events. Accounting events are multidimensional as they have three dimensions—debit, credit, and size, denoted as D, C, and S for short (see Sect. 1.1, Chap. 1).

Type 1 Events: A Single Account (Entry) on Both Sides

To see how this type of accounting events can be examined in the three-dimensional vector space, refer back to the type 1 accounting event (seen before)—the sale of goods in cash. Figure 6.7 presents the accounting event in the three-dimensional vector space. The shaded cube is the DE cube of the accounting event. Since the accounting event increases the debit and credit accounts alike, the DE cube lies in the first octant (OI) of the three-dimensional vector space. As coordinate vectors, **d**, **c**, and **q** are mutually orthogonal and have the same magnitude (|**d**| = |**c**| = |**q**| = 100) under the double entry principle. As a set of basis vectors, they generate the DE vector **D** and span the DE cube of the accounting event. **D** has an angle of 35.2645° with the bottom face of the DE cube. Thus, it has a slope of 0.7071. Shown below are those vectors above.

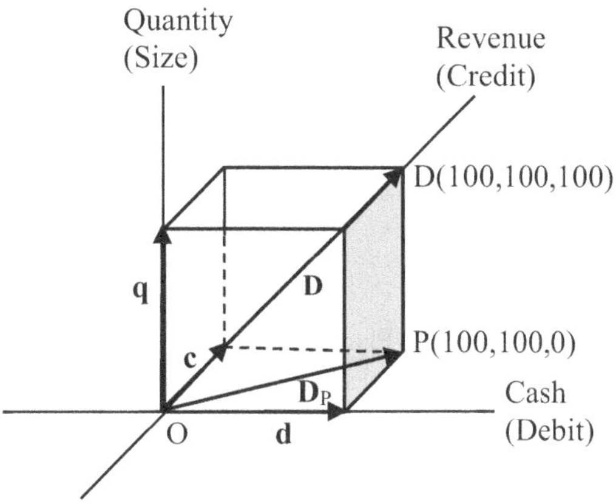

Fig. 6.7 Presenting a type 1 accounting event in the three-dimensional vector space

$$\begin{aligned}
\mathbf{d} &= (100, 0, 0)\\
\mathbf{c} &= (0, 100, 0)\\
\mathbf{q} &= (0, 0, 100)\\
\mathbf{D} &= \mathbf{d} + \mathbf{c} + \mathbf{q} = (100, 100, 100)\\
\mathbf{D}_P &= \mathbf{d} + \mathbf{c} = (100, 100, 0)
\end{aligned}$$

As seen, **D** is a linear combination of the coordinate vectors, each scaled by a factor of 1. The vector $\mathbf{D}_P$ is a perpendicular projection of **D** onto the bottom face of the DE cube. It is the DE vector of the accounting event in the debit-credit plane of the three-dimensional vector space.

The size (q) of the accounting event in monetary terms is \$100, which is equal to the magnitude of **q**. If measured geometrically in the three-dimensional vector space, it is equal to the magnitude |**D**| or the size of the DE cube. Each is computed below.

$$\begin{aligned}
|\mathbf{D}| &= \left(OP^2 + DP^2\right)^{1/2} = \left[\left(|\mathbf{d}|^2 + |\mathbf{c}|^2\right) + |\mathbf{q}|^2\right]^{1/2}\\
&= 100\sqrt{3} = 173.2051
\end{aligned}$$

and

$$Q_S = |\mathbf{d}||\mathbf{c}||\mathbf{q}| = d.c.q = 100^3 vu^3$$

The magnitude |**D**| is computed based on the algebraic definition of the magnitude of vectors in Sect. 6.1 before or according to the Pythagorean theorem. As a geometric measure of the size of the accounting event, |**D**| is equivalent to the line-based measure Q_L of the size of the event in Sect. 4.4, Chap. 4. And Q_S is equivalent to the space-based measure Q_S of the size of the event in Sect. 4.2, Chap. 4.

The two DE vectors $\mathbf{D}$ and $\mathbf{D}_P$ define the right triangle OPD. Given this triangle, $|\mathbf{D}|$ can also be computed according to the law of cosine or the law of sine as follows:

$$\begin{aligned}|\mathbf{D}| &= |OP|/\text{cosine }\theta \text{ or } |DP|/\text{sine }\theta \\ &= (|\mathbf{d}|/\text{cosine } 45^{o})/\text{cosine }\theta \text{ or } |\mathbf{q}|/\text{sine }\theta \\ &= (100/0.70710678)/\text{cosine }\theta \text{ or } 100/\text{sine }\theta \\ &= 173.2046\end{aligned}$$

where |OP| and |DP| are the length of the lines OP and DP respectively, and θ is the angle between $\mathbf{D}$ and $\mathbf{D}_P$ ($\theta \approx 35.2645°$).

As a geometric measure of the size of the accounting event, $|\mathbf{D}|$ can be converted to the monetary size of the event as follows:

$$\begin{aligned}q = |\mathbf{q}| &= \left[|\mathbf{D}|^2 - \left(|\mathbf{d}|^2 + |\mathbf{c}|^2\right)\right]^{1/2} \\ &= \left[173.2051^2 - \left(100^2 + 100^2\right)\right]^{1/2} = 100\end{aligned}$$

or

$$q = |\mathbf{D}|\text{sine } 35.2645^{o} = (173.2051)(0.577352) = 100$$

Type 2 Events: Multiple Accounts (Entries) on the Debit Side

To see how accounting events of this type can be examined in the three-dimensional vector space, refer back to the type 2 accounting event (seen before)—the sale of goods partly in cash and partly on account. This event has two component transactions. Figure 6.8 presents the accounting event in the three-dimensional vector space. The unshaded DE cube represents the event, and the small shaded DE cube shown represents the cash part of the event. The small DE cube of the account-receivable part of the accounting event is not shown (to keep the figure simple).[5] Since the accounting event increases the debit and credit accounts alike, the DE cube lies in OI.

Given the two component transactions, each coordinate vector is divided into two discrete vectors of order 3, one for the cash part of the accounting event and the other for the account-receivable part: $\mathbf{d}_1 = (60, 0, 0)$, $\mathbf{d}_2 = (100–60, 0, 0) = (40, 0, 0)$, $\mathbf{c}_1 = (0, 60, 0)$, $\mathbf{c}_2 = (0, 100–60, 0) = (0, 40, 0)$, $\mathbf{q}_1 = (0, 0, 60)$, and $\mathbf{q}_2 = (0, 0, 100–60) = (0, 0, 40)$. Thus, the DE vector $\mathbf{D}$ of the accounting event is divided accordingly: $\mathbf{D}_1$ for the cash part and $\mathbf{D}_2$ for the account-receivable part. $\mathbf{D}_1$ and $\mathbf{D}_2$ are collinear and lie in the same octant of the vector space. They can be linearly

[5]If shown, it would be in the upper-right corner of the large DE cube. It is defined by the three coordinate vectors $\mathbf{d}_2$, $\mathbf{c}_2$, and $\mathbf{q}_2$.

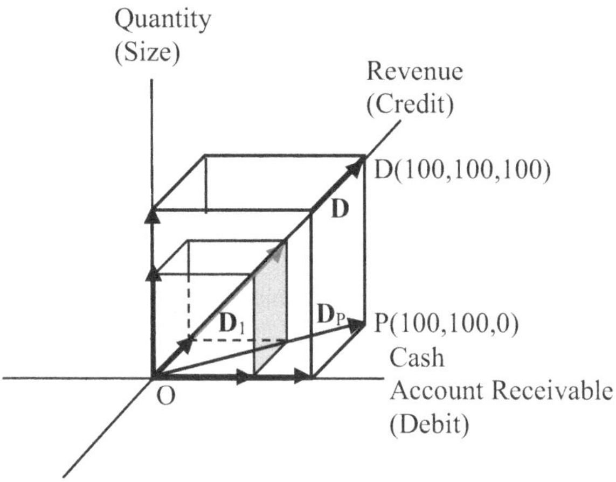

Fig. 6.8 Presenting a type 2 accounting event in the three-dimensional vector space

added to get **D** as a single vector. **d**, **c**, and **q** are mutually orthogonal and have the same magnitude. As a set of basis vectors, the three coordinate vectors generate **D** and span the large unshaded DE cube. Shown below are all those vectors of the accounting event.

$$\begin{aligned}
\mathbf{d} &= \mathbf{d}_1 + \mathbf{d}_2 = (60 + 40, 0, 0) = (100, 0, 0)\\
\mathbf{c} &= \mathbf{c}_1 + \mathbf{c}_2 = (0, 60 + 40, 0) = (0, 100, 0)\\
\mathbf{q} &= \mathbf{q}_1 + \mathbf{q}_2 = (0, 0, 60 + 40) = (0, 0, 100)\\
\mathbf{D}_1 &= (\mathbf{d}_1 + \mathbf{c}_1 + \mathbf{q}_1) = (60, 60, 60)\\
\mathbf{D}_2 &= (\mathbf{d}_2 + \mathbf{c}_2 + \mathbf{q}_2) = (40, 40, 40)\\
\mathbf{D} &= \mathbf{D}_1 + \mathbf{D}_2 = (\mathbf{d} + \mathbf{c} + \mathbf{q}) = (100, 100, 100)\\
\mathbf{D}_\mathrm{P} &= \mathbf{d} + \mathbf{c} = (100, 100, 0)
\end{aligned}$$

The size of the accounting event in monetary terms is \$100, equal to the magnitude of **q** or $|\mathbf{q}_1| + |\mathbf{q}_2|$. If measured geometrically in the vector space, it equals the magnitude of **D** or the size of the DE cube. Each is computed below.

$$\begin{aligned}
|\mathbf{D}| &= |\mathbf{D}_1| + |\mathbf{D}_2|\\
&= \left[\left(|\mathbf{d}_1|^2 + |\mathbf{c}_1|^2\right) + |\mathbf{q}_1|^2\right]^{1/2} + \left[\left(|\mathbf{d}_2|^2 + |\mathbf{c}_2|^2\right) + |\mathbf{q}_2|^2\right]^{1/2}\\
&= \left(60^2 + 60^2 + 60^2\right)^{1/2} + \left(40^2 + 40^2 + 40^2\right)^{1/2}\\
&= 173.2051
\end{aligned}$$

and

$$\mathrm{Q_S} = |\mathbf{d}||\mathbf{c}||\mathbf{q}| = \mathrm{d.c.q} = 100^3\mathrm{vu}^3$$

In magnitude, |**D**| is equal to the line-based measure Q_L of the size of the accounting event in Sect. 4.4, Chap. 4. And Q_S is equal to the space-based measure Q_S of the size of the event in Sect. 4.2, Chap. 4.

Together, the DE vectors **D** and $\mathbf{D}_P$ define the right triangle OPD. Given this triangle, the line-based measure |**D**| of the size of the accounting event is also computed as

$$
\begin{aligned}
|\mathbf{D}| &= |\mathrm{OP}|/\text{cosine }\theta \text{ or } |\mathrm{DP}|/\text{sine }\theta \\
&= [(|\mathbf{d}_1|/\text{cosine }45^{\circ}) + (|\mathbf{d}_2|/\text{cosine }45^{\circ})]/\text{cosine }\theta \\
&= [(60/\text{cosine }45^{\circ}) + (40/\text{cosine }45^{\circ})]/\text{cosine }\theta \\
&= [(60/0.707107) + (40/0.707107)]/\text{cosine }\theta \\
&= 173.2046
\end{aligned}
$$

or

$$
\begin{aligned}
|\mathbf{D}| &= (|\mathbf{q_1}|/\text{sine }\theta + |\mathbf{q_2}|/\text{sine }\theta) = (60 + 40)/0.577352 \\
&= 173.2046
\end{aligned}
$$

where $\theta \approx 35.2645°$.

Since the alternative geometric measures of the size of the accounting event are equivalent, they can be converted to the same monetary size ($100) of the event as shown below.

$$
\begin{aligned}
q &= |\mathbf{q}| = \left[|\mathbf{D}|^2 - (|\mathbf{d}_1| + |\mathbf{d}_2|)^2 - (|\mathbf{c}_1| + |\mathbf{c}_2|)^2\right]^{1/2} \\
&= \left[173.2051^2 - (60 + 40)^2 - (60 + 40)^2\right]^{1/2} \\
&= 100
\end{aligned}
$$

or

$$
\begin{aligned}
q &= |\mathbf{q}_1| + |\mathbf{q}_2| \\
&= \left[|\mathbf{D}_1|^2 - \left(|\mathbf{d}_1|^2 + |\mathbf{c}_1|^2\right)\right]^{1/2} + \left[|\mathbf{D}_2|^2 - \left(|\mathbf{d}_2|^2 + |\mathbf{c}_2|^2\right)\right]^{1/2} \\
&= \left[103.9231^2 - (60^2 + 60^2)\right]^{1/2} + \left[69.2820^2 - (40^2 + 40^2)\right]^{1/2} \\
&= 100
\end{aligned}
$$

or

$$q = |\mathbf{D}|\text{sine }35.2645^{\circ} = 100$$

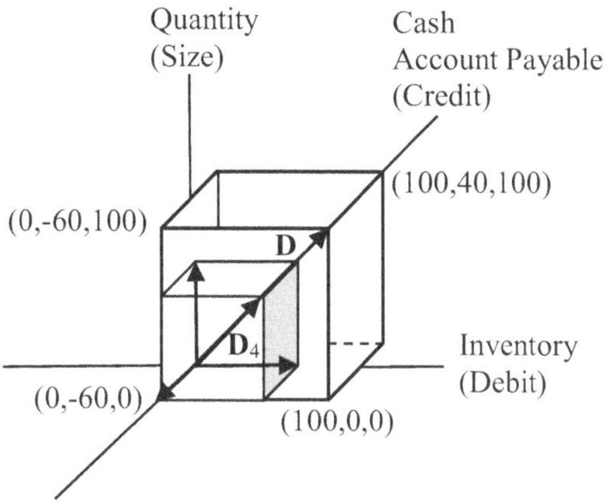

Fig. 6.9 Presenting a type 3 accounting event in the three-dimensional vector space

Type 3 Events: Multiple Accounts (Entries) on the Credit Side

To examine this type of accounting events in the three-dimensional vector space, consider the type 3 event (seen before)—that a firm purchases goods, partly in cash and partly on account. The inventory asset \$100 is debited as an increase, and the cash paid is credited as a decrease. The event has two component transactions, one for the cash part and the other for the account-payable part. Figure 6.9 presents the accounting event in the three-dimensional vector space. The unshaded cube is the DE cube of the accounting event. It lies across OI and OIV of the three-dimensional vector space since the event increases the account payable and decreases the cash, both on the credit side of the journal entry. The small shaded cube in OIV is the DE cube of the cash part of the event.[6] It is defined by the three coordinate vectors: $\mathbf{d}_1$, $\mathbf{a}$, and $\mathbf{q}_1$. The corresponding coordinate vectors of the account-payable part of the accounting event are $\mathbf{d}_2$, $\mathbf{c}$, and $\mathbf{q}_2$. As a set of basis vectors, $\mathbf{d}_1$, $\mathbf{a}$, and $\mathbf{q}_1$ ($\mathbf{d}_2$, $\mathbf{c}$, and $\mathbf{q}_2$) generate $\mathbf{D}_4$ ($\mathbf{D}_1$), and span the small DE cube in OIV (OI). Shown below are all those vectors relating to the accounting event.

$$\begin{aligned}
\mathbf{d} &= \mathbf{d}_1 + \mathbf{d}_2 = (60, 0, 0) + (40, 0, 0) = (100, 0, 0) \\
\mathbf{c} &= (0, 40, 0) \\
\mathbf{a} &= (0, -60, 0) \\
\mathbf{q} &= \mathbf{q}_1 + \mathbf{q}_2 = (0, 0, 60) + (0, 0, 40) = (0, 0, 100) \\
\mathbf{D}_1 &= \mathbf{d}_2 + \mathbf{c} + \mathbf{q}_2 = (40, 40, 40) \\
\mathbf{D}_4 &= \mathbf{d}_1 + \mathbf{a} + \mathbf{q}_1 = (60, -60, 60)
\end{aligned}$$

[6] The small DE cube of the account-payable part of the accounting event is not shown to keep the figure simple. It is defined by $\mathbf{d}_2$, $\mathbf{c}$, and $\mathbf{q}_2$. It lies in OI, behind the small shaded DE cube shown.

where the subscripts 1 and 4 of $\mathbf{D}_1$ and $\mathbf{D}_4$ denote OI and OIV, respectively. Since $\mathbf{D}_1$ and $\mathbf{D}_4$ are not collinear, they cannot be linearly added to get $\mathbf{D}$ as a single vector. Thus, the accounting event is jointly represented by $\mathbf{D}_1$ and $\mathbf{D}_4$.

The size (quantity) of the accounting event measured in monetary terms is \$100, which is equal to the magnitude of $\mathbf{q}_1$ and $\mathbf{q}_2$ combined. If measured geometrically in the vector space, it is equal to the combined magnitude of $\mathbf{D}_1$ and $\mathbf{D}_4$ or the size of the DE cube of the event. Each is computed below.

$$\begin{aligned}|\mathbf{D}| &= |\mathbf{D}_1| + |\mathbf{D}_4| \\ &= \left[\left(|\mathbf{d}_2|^2 + |\mathbf{c}|^2\right) + |\mathbf{q}_2|^2\right]^{1/2} + \left[\left(|\mathbf{d}_1|^2 + |\mathbf{a}|^2\right) + |\mathbf{q}_1|^2\right]^{1/2} \\ &= \left(40^2 + 40^2 + 40^2\right)^{1/2} + \left(60^2 + |-60|^2 + 60^2\right)^{1/2} \\ &= 173.2051\end{aligned}$$

and

$$Q_S = |\mathbf{d}|(|\mathbf{c}| + |\mathbf{a}|)|\mathbf{q}| = 100^3\mathrm{vu}^3$$

While $\mathbf{D}_1$ and $\mathbf{D}_4$ cannot be linearly added to get $\mathbf{D}$ as a single vector, their magnitudes can be. $|\mathbf{D}|$ is equivalent to the line-based measure Q_L of the size of the accounting event in Sect. 4.4, Chap. 4. And Q_S is equivalent to the space-based measure Q_S of the size of the event in Sect. 4.2, Chap. 4.

The vectors $\mathbf{D}_1$ and $\mathbf{D}_4$ each, along with the related debit and credit vectors, define a right triangle (not shown in the figure), similar to the right triangle OPD in the previous figure. Given the right triangle, $|\mathbf{D}|$ can also be computed according to the law of cosine as

$$\begin{aligned}|\mathbf{D}| &= |\mathbf{D}_4| + |\mathbf{D}_1| \\ &= [(|\mathbf{d}_1|/\text{cosine } 45^\circ) + (|\mathbf{d}_2|/\text{cosine } 45^\circ)]/\text{cosine } \theta \\ &= [(60/\text{cosine } 45^\circ) + (40/\text{cosine } 45^\circ)]/\text{cosine } \theta \\ &= [(60/0.707107) + (40/0.707107)]/\text{cosine } \theta \\ &= 173.2046\end{aligned}$$

where $\theta \approx 35.2645^\circ$. The line-based measure of the size of the accounting event can also be computed according to the law of sine as follows:

$$\begin{aligned}|\mathbf{D}| &= (|\mathbf{q}_1|/\text{sine } \theta) + (|\mathbf{q}_2|/\text{sine } \theta) \\ &= (60/0.577352) + (40/0.577352) \\ &= 173.2046\end{aligned}$$

Since the alternative geometric measures above of the size of the accounting event are equivalent, they can be converted to the same monetary size of the event:

$$\begin{aligned} q &= |\mathbf{q}| = \left[|\mathbf{D}|^2 - (|\mathbf{d}_1| + |\mathbf{d}_2|)^2 - (|\mathbf{c}| + |\mathbf{a}|)^2\right]^{1/2} \\ &= \left[173.20508^2 - (60 + 40)^2 - (40 + |-60|)^2\right]^{1/2} \\ &= 100 \end{aligned}$$

or

$$\begin{aligned} q &= |\mathbf{q}_1| + |\mathbf{q}_2| \\ &= \left[|\mathbf{D}_1|^2 - \left(|\mathbf{d}_2|^2 + |\mathbf{c}|^2\right)\right]^{1/2} + \left[|\mathbf{D}_4|^2 - \left(|\mathbf{d}_1|^2 + |\mathbf{a}|^2\right)\right]^{1/2} \\ &= \left[69.282^2 - (40^2 + 40^2)\right]^{1/2} + \left[103.923^2 - \left(60^2 + |-60|^2\right)\right]^{1/2} \\ &= 100 \end{aligned}$$

or

$$q = |\mathbf{D}| \text{ sine } 35.2645^{\circ} = 100$$

Type 4 Events: Multiple Accounts (Entries) on Both Sides

To see how this type of accounting events can be presented in the three-dimensional vector space, refer back to the type 4 accounting event (seen before)—that a firm recognizes an interest expense on a bond liability, pays the accrued interest in cash on the bond liability, and amortizes a bond discount, all as one transaction. This event has three component transactions. Figure 6.10 presents the event in the three-dimensional vector space. The large unshaded cube is the DE cube of the accounting event. Given the three component transactions, there are three small shaded DE cubes, one for each component transaction. The small DE cube in OI represents the increase in the bond liability due to the amortization of the bond discount \$100 (not seen in the figure). The small shaded DE cube in OIII represents the payment in cash of the past-period interest \$120, and the small shaded DE cube in OIV represents the payment of the current-period interest \$230. The small shaded cube in OI is defined by $\mathbf{d}_1$, $\mathbf{c}$, and $\mathbf{q}_1$, the one in OIII by $\mathbf{b}$, $\mathbf{a}_1$, and $\mathbf{q}_2$, and the one in OIV by $\mathbf{d}_2$, $\mathbf{a}_2$, and $\mathbf{q}_3$. There is no small DE cube in OII since the accounting event has no component transaction that has a debit account as a decrease and a credit account as an increase. $\mathbf{D}_1$, $\mathbf{D}_3$, and $\mathbf{D}_4$ are the DE vectors of the three respective component transactions. Together, they represent the accounting event as a whole.

Shown below are all those vectors for the accounting event.

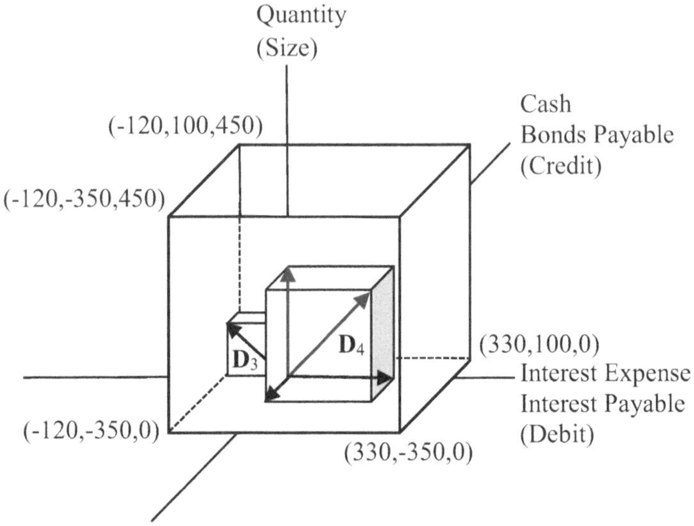

Fig. 6.10 Presenting a type 4 accounting event in the three-dimensional vector space

$$
\begin{aligned}
\mathbf{d} &= \mathbf{d}_1 + \mathbf{d}_2 = (100, 0, 0) + (230, 0, 0) = (330, 0, 0) \\
\mathbf{c} &= (0, 100, 0) \\
\mathbf{b} &= (-120, 0, 0) \\
\mathbf{a} &= \mathbf{a}_1 + \mathbf{a}_2 = (0, -120, 0) + (0, -230, 0) = (0, -350, 0) \\
\mathbf{q} &= \mathbf{q}_1 + \mathbf{q}_2 + \mathbf{q}_3 = (0, 0, 100) + (0, 0, 120) + (0, 0, 230) \\
&= (0, 0, 450) \\
\mathbf{D}_1 &= \mathbf{d}_1 + \mathbf{c} + \mathbf{q}_1 = (100, 100, 100). \\
\mathbf{D}_3 &= \mathbf{b} + \mathbf{a}_1 + \mathbf{q}_2 = (-120, -120, 120) \\
\mathbf{D}_4 &= \mathbf{d}_2 + \mathbf{a}_2 + \mathbf{q}_3 = (230, -230, 230)
\end{aligned}
$$

where the subscripts 1, 3, and 4 of $\mathbf{D}_1$, $\mathbf{D}_3$, and $\mathbf{D}_4$ denote the octants of the three-dimensional space where these vectors lie. Since the three DE vectors are not collinear, they cannot be added to get $\mathbf{D}$ as a single DE vector.

The size (quantity) of the accounting event in monetary terms is \$450, equal to the magnitude of $\mathbf{q}$. If measured geometrically in the vector space, it is equal to the combined length of $\mathbf{D}_1$, $\mathbf{D}_3$, and $\mathbf{D}_4$ or the size of the large unshaded DE cube. Each is computed below.

$$\begin{aligned}|\mathbf{D}| &= |\mathbf{D}_1| + |\mathbf{D}_3| + |\mathbf{D}_4|\\ &= \left[\left(|\mathbf{d}_1|^2 + |\mathbf{c}|^2\right) + |\mathbf{q}_1|^2\right]^{1/2} + \left[\left(|\mathbf{b}|^2 + |\mathbf{a}_1|^2\right) + |\mathbf{q}_2|^2\right]^{1/2}\\ &\quad + \left[\left(|\mathbf{d}_2|^2 + |\mathbf{a}_2|^2\right) + |\mathbf{q}_3|^2\right]^{1/2}\\ &= 100\sqrt{3} + 120\sqrt{3} + 230\sqrt{3}\\ &= 779.42286\end{aligned}$$

and

$$\begin{aligned}Q_S &= (|\mathbf{d}_1| + |\mathbf{d}_2| + |\mathbf{b}|)(|\mathbf{c}| + |\mathbf{a}_1| + |\mathbf{a}_2|)(|\mathbf{q}_1| + |\mathbf{q}_2| + |\mathbf{q}_3|)\\ &= 450^3\,\text{vu}^3\end{aligned}$$

Although $\mathbf{D}_1$, $\mathbf{D}_3$, and $\mathbf{D}_4$ cannot be linearly added to get $\mathbf{D}$ as a single vector, their magnitudes can be. As a measure of the size of the accounting event, $|\mathbf{D}|$ is equivalent to the line-based measure Q_L of the size of the same event in Sect. 4.4, Chap. 4. And Q_S is equivalent to the space-based measure Q_S of the size of the event in Sect. 4.2, Chap. 4.

The line-based measure $|\mathbf{D}|$ of the size of the accounting event can also be computed according to the law of cosine or the law of sine as follows:

$$\begin{aligned}|\mathbf{D}| &= |\mathbf{D}_1| + |\mathbf{D}_3| + |\mathbf{D}_4|\\ &= (|\mathbf{d}_1|/\text{cosine } 45^\circ)/\text{cosine } \theta + (|\mathbf{b}|/\text{cosine } 45^\circ)/\text{cosine } \theta\\ &\quad + (|\mathbf{d}_2|/\text{cosine } 45^\circ)/\text{cosine } \theta\\ &= (100/\text{cosine } 45^\circ)/\text{cosine } \theta + (|-120|/\text{cosine } 45^\circ)/\text{cosine } \theta\\ &\quad + (230/\text{cosine } 45^\circ)/\text{cosine } \theta\\ &= 779.4242\end{aligned}$$

or

$$\begin{aligned}|\mathbf{D}| &= |\mathbf{q}_1|/\text{sine } \theta + |\mathbf{q}_2|/\text{sine } \theta + |\mathbf{q}_3|/\text{sine } \theta\\ &= (|\mathbf{q}_1| + |\mathbf{q}_2| + |\mathbf{q}_3|)/\text{sine } \theta = (100 + 120 + 230)/0.57735\\ &= 779.4205\end{aligned}$$

where $\theta \approx 35.2645^\circ$.

Since the alternative geometric measures of the size of the accounting event are equivalent, they can be converted to the same monetary size of the event as follows:

$$|\mathbf{q}| = |\mathbf{q}_1| + |\mathbf{q}_2| + |\mathbf{q}_3| = 100 + 120 + 230 = 450$$

or

$$
\begin{aligned}
|\mathbf{q}| &= \left[|\mathbf{D}_1|^2 - \left(|\mathbf{d}_1|^2 + |\mathbf{c}|^2\right)\right]^{1/2} + \left[|\mathbf{D}_3|^2 - \left(|\mathbf{b}|^2 + |\mathbf{a}_1|^2\right)\right]^{1/2} \\
&\quad + \left[|\mathbf{D}_4|^2 - \left(|\mathbf{d}_2|^2 + |\mathbf{a}_2|^2\right)\right]^{1/2} \\
&= \left[(3)\left(100^2\right) - \left(100^2 + 100^2\right)\right]^{1/2} \\
&\quad + \left[(3)|-120|^2 - \left(|-120|^2 + |-120|^2\right)\right]^{1/2} \\
&\quad + \left[(3)\left(230^2\right) - \left(230^2 + 230^2\right)\right]^{1/2} = 450
\end{aligned}
$$

or

$$
|\mathbf{q}| = |\mathbf{D}| \operatorname{sine} \theta = 779.4208 \operatorname{sine} \theta = 450
$$

6.5 The Forces and Power of Accounting Events in Vector Spaces

The Forces of Accounting Events in Vector Spaces

The concept *force* is closely related to the concept *vector* as they both have common attributes—magnitude and direction. Because of this, a force is often called a vector force, and a vector is called a force vector.[7] As such, the magnitude of a force can be measured by the length of the force vector. It is independent of the direction of the force vector. Having both magnitude and direction, a force vector is written as an array of two real numbers—one for its magnitude and the other for its direction—in a two-dimensional vector space. In a three-dimensional vector space, it is written as an array of three real numbers.

Accounting events have a force in that they affect the sources and uses of the capital resources of a business entity (firm) and their relationship as shown by the fundamental accounting model (see Sect. 5.1, Chap. 5). As a force is a vector in a vector space, so is the force of accounting events. The force of accounting events is nonphysical as accounting events are nonphysical. It is generated by the debit and credit forces that have the same magnitude and act in perpendicular directions under the double entry principle. In this regard, it is an equiforce. Hereafter, the vector representing this force is called the equiforce vector of accounting events, denoted as **D**. It is generated by the debit and credit force vectors **d** and **c** in the two-dimensional (debit-credit) vector space and by the debit, credit, and size force vectors **d**, **c**, and **q** in the three-dimensional (debit-credit-size) vector space. For a given accounting event, **d** and **c** as a pair, and **d**, **c**, and **q** as a trio, have the same magnitude and act together in perpendicular directions under the double entry principle. And the

[7] For the concept *force* versus *vector*, see https://en.wikipedia.org/wiki/Force (August 2020).

equiforce vector **D** represents the relationship between those component vectors and reflects the duality and balance features of the accounting event.

The equiforce vector of accounting events is written as a linear combination of the debit and credit force vectors or the debit, credit, and size force vectors of accounting events, depending on the dimensionality of the vector space where those vectors lie:

$$\mathbf{D} = \mathbf{d} + \mathbf{c} \quad \text{(in two-dimensional space)}$$

and

$$= \mathbf{d} + \mathbf{c} + \mathbf{q} \quad \text{(in three-dimensional space)}$$

With **D** included, every accounting event has three or four force vectors, **d**, **c**, and **D** or **d**, **c**, **q**, and **D**. They are all position vectors of order 2 or order 3 and start at the point of origin (zero quantity) of the vector space. For a given accounting event, they co-exist and act together according to the double entry principle. Thus, they are interrelated and dependent even though they point in different directions.

The equiforce vector **D** above concerns accounting events of type 1, the simplest type possible, which have no component transactions. If accounting events are of type 2, 3, or 4, they have component transactions. Thus, **D** is not one continuous force vector but is divided into multiple discrete force vectors. Each of these discrete force vectors represents the equiforce of a component transaction. The discrete force vectors may or may not point in the same direction since the component transactions of such accounting events may or may not change the debit and credit accounts in the same direction. If not, they cannot be linearly added to get **D** as a single equiforce vector. Thus, the equiforce of the accounting events is represented by a set of the discrete equiforce vectors relating to their component transactions.

Besides their equiforce vectors, type 2, 3, or 4 accounting events also have a set of vectors for interaction forces. These interaction forces are generated by the cross interaction of the debit (credit) force of one component transaction and the credit (debit) force of another. As with forces in general, they are also considered vectors and referred to as interaction force vectors hereafter. While the equiforce vector exists for every accounting event, interaction force vectors exist only for accounting events that have component transactions. Also, when an accounting event has an interaction force vector, it has two such vectors at a minimum since it has a minimum of two component transactions. Unlike the equiforce vector of accounting events, however, an interaction force vector does not reflect the duality and balance features of the accounting event or a component transaction. This is because it is generated not by a particular component transaction but by the *cross* interaction of the debit (credit) force of one component transaction and the credit (debit) force of another.

In the two-dimensional (debit-credit) vector space, the debit and credit force vectors of accounting events are shown by an arrow along the respective debit and credit coordinates. The dual force vectors are orthogonal ($\theta = 90°$), have the same magnitude, and act together according to the double entry principle. Thus, the vector space spanned by them is a square (area)—a DE square. The diagonal of this square is equal to the equiforce vector **D** of the accounting events. **D** has an angle of 45° with the debit and credit coordinates. In the three-dimensional (debit-credit-size) vector space, there is one more force vector—a size force vector—for every accounting event. The three (debit, credit, and size) force vectors are orthogonal and have the same magnitude. They are interrelated and act together according to the double entry principle. In this sense, they are mutually dependent even though they point in perpendicular directions. Together, they generate **D** and span a vector space—the DE cube of the accounting event. And **D** is equal to the diagonal of the DE cube. As an equiforce vector, **D** has an angle of 35.2645° with the bottom face of the cube and has a slope of 0.7071.

If an accounting event is of type 1 and increases the debit and credit accounts alike, the equiforce vector **D** lies in the first quadrant (QI) of the two-dimensional vector space and in the first octant (OI) of the three-dimensional vector space. If an accounting event is of type 1 but decreases the account on each side, **D** lies in QIII (OIII). If neither is the case, it lies in QII (OII) or QIV (OIV). If an accounting event is of type 2, 3, or 4, **D** is not one continuous force vector but is divided into multiple discrete equiforce vectors. These discrete equiforce vectors may or may not have the same magnitude. Nor do they need to point in the same direction. Thus, they do not always lie in the same quadrant or octant of the multidimensional vector space.

The Power of Accounting Events in Vector Spaces

Power is generated by a force, not vice versa. It is increasing with the magnitude of its underlying force but independent of the direction of the force. Unlike a force, it has magnitude but no direction and, hence, is considered an absolute scalar quantity. Also, it is a multiplicative, not an additive, concept. Given the multiplicative nature of power, the space-based presentation of power and the space-based measurement of its magnitude are more reasonable than are the line- or point-based presentation of it and the line- or point-based measurement of its magnitude (see Sect. 5.2, Chap. 5).

Accounting events have a force and, hence, power. This power is generated by the forces of the accounting events. Geometrically, the power of an accounting event is considered to represent the strength of its capability to stretch its equiforce vector along with its debit and credit (and size) force vectors, thereby generating a vector space representing the event. This space is the DE vector space of the accounting event. The larger this vector space, the greater is the power of the event, all else equal. The DE vector space is a DE square in the two-dimensional vector space and its size, not diagonal, is a geometric measure of the power of the event. In the three-dimensional vector space, the power of an accounting event is also defined similarly, except that the force vectors involved are of order 3. It represents the strength of the capability of the accounting event to stretch the equiforce vector along with the related debit, credit, and size force vectors, thereby generating a cubic space. This space is the DE cube of the accounting event. And the size of the DE cube is a geometric measure of the power of the event. Again, the larger the size of the cube, the greater is the power of the accounting event.

Given the above, the power of an accounting event can be measured by the size of its DE square or DE cube in the multidimensional vector space as follows:

$$P = |\mathbf{d}||\mathbf{c}| = d.c \quad \text{(in two-dimensional space)}$$

or

$$P = |\mathbf{d}||\mathbf{c}||\mathbf{q}| = d.c.q \quad \text{(in three-dimensional space)}$$

As a space-based measure of power, P is equal to the size of the DE square or DE cube of the accounting event. When the power of accounting events is measured this way, it is consistent with the notion that size matters when it comes to power. That is, the larger the (monetary) size of an accounting event, the larger the DE square or DE cube, the greater is the power of the event, all else equal.

The space-based measure of power shown above concerns an accounting event of type 1, which has no component transactions. If an accounting event is of type 2, 3, or 4, thus having component transactions, the debit and credit force vectors of the event are divided into multiple discrete force vectors. So are the equiforce vector and DE vector space respectively of the event. As such, the magnitude of the power of the accounting event is not one continuous magnitude but a multitude of discrete magnitudes. Yet it is still measured in the same way, namely by the size of the DE vector space of the event. In the two-dimensional vector space, the size of this vector space is equal to the combined size of the component DE squares and the related rectangles defined by the discrete debit and credit force vectors of the component transactions. In the three-dimensional vector space, it is equal to the combined size of the component DE cubes and the related rectangular cuboids defined by the discrete debit, credit, and size force vectors of those component transactions. The discrete equiforce vectors of the accounting event may or may not be collinear. Nor do they need to lie in the same quadrant or octant of the multidimensional vector space. Yet the power of the accounting event can be measured by the combined size of those discrete vector spaces since power is an absolute scalar quantity. Measuring the power of type 2, 3, and 4 accounting events as said above is consistent with the "whole-parts" relation—that the "whole" is the same as the sum of its "parts."

The discussions above of the forces and power of accounting events can be illustrated in two- and three-dimensional vector spaces with specific transaction examples.

The Force Vectors and Power of Type 1 Accounting Events

To see how the forces and power of accounting events of this type can be examined in vector spaces, refer back to the type 1 accounting event in Fig. 6.3. The debit and credit vectors of the event there are now the debit and credit force vectors: $\mathbf{d} = (100, 0)$ and $\mathbf{c} = (0, 100)$. The DE vector $\mathbf{D}$ there is now the equiforce vector of the event. And the power of the event is measured by the size of the shaded DE vector space (DE square), which is spanned by **d** and **c** as a pair of basis vectors. Both of these are shown below.

$$\mathbf{D} = \mathbf{d} + \mathbf{c} = (100, 100)$$

and

$$\mathrm{P} = |\mathbf{d}||\mathbf{c}| = \mathrm{d.c} = 100^2 \mathrm{au}^2$$

The space-based measure P of the power of the accounting event is equivalent to the space-based measure of the power of the same event in Sect. 5.3, Chap. 5. It is considered a geometric presentation of the extent of the impact of the accounting event on both the cash account and the revenue account affected by the event. That is, it is a geometric measure of the strength of the capability of the accounting event to stretch its equiforce vector along with the debit and credit force vectors in the two-dimensional vector space, thereby generating the DE square whose size is 100^2 au^2.

In a three-dimensional vector space, the accounting event has another force vector—the size force vector **q**, in addition to **d**, **c**, and **D**. All these vectors are of order 3: **d** = (100, 0, 0), **c** = (0, 100, 0), **q** = (0, 0, 100), and **D** = (100, 100, 100). **D** is defined as a linear combination of the three force vectors as shown below. It represents the equiforce of the accounting event. As a set of basis vectors, **d**, **c**, and **q** generate **D** and span the DE cube of the accounting event, like the one in Fig. 6.7. The power of the accounting event is measured by the size of the DE cube.

$$\mathbf{D} = \mathbf{d} + \mathbf{c} + \mathbf{q} = (100, 100, 100)$$

and

$$\mathrm{P} = |\mathbf{d}||\mathbf{c}||\mathbf{q}| = 100^3 \mathrm{vu}^3$$

The space-based measure P of power is equivalent to the space-based measure of the power of the same event in Sect. 5.3, Chap. 5. It is a geometric presentation of the extent of the impact of the accounting event on the debit and credit accounts directly affected by the event. That is, it is a geometric measure of the strength of the capability of the accounting event to stretch the equiforce vector **D** along with the force vectors **d**, **c**, and **q**, thereby creating the cubic DE vector space whose size is equal to 100^3 vu^3.

The Force Vectors and Power of Type 4 Accounting Events

To see how the forces and power of accounting events of this type can be examined in the two-dimensional vector space, refer back to the accounting event in Fig. 6.6. The event has two accounts on both debit and credit sides and three component transactions—the payment of the unpaid interest \$120, the payment of the current-period interest \$230, and the amortization of the bond discount \$100. Thus, it has two force vectors on the debit and credit sides, respectively: $\mathbf{d} = \mathbf{d}_1 + \mathbf{d}_2 = (100, 0) + (230, 0) = (330, 0)$ and $\mathbf{b} = (-120, 0)$ on the debit side, and $\mathbf{c} = (0, 100)$ and $\mathbf{a} = \mathbf{a}_1 + \mathbf{a}_2 = (0, -120) + (0, -230) = (0, -350)$ on the credit side. And $|\mathbf{d}_1| + |\mathbf{d}_2| + |\mathbf{b}| = |\mathbf{c}| + |\mathbf{a}_1| + |\mathbf{a}_2|$ under the double entry principle. Given the three component transactions, the accounting event has three discrete equiforce vectors, written as

$$\mathbf{D}_1 = \mathbf{d}_1 + \mathbf{c} = (100, 100)$$

$$\mathbf{D}_3 = \mathbf{b} + \mathbf{a}_1 = (-120, -120)$$

$$\mathbf{D}_4 = \mathbf{d}_2 + \mathbf{a}_2 = (230, -230)$$

The equiforce vector $\mathbf{D}_1$, which lies in QI of the vector space, represents the equiforce of the first component transaction—the increase in the bond liability due to the amortization of bond discount. $\mathbf{D}_3$ in QIII represents the equiforce of the second component transaction—the decrease in the interest payable as a liability. And $\mathbf{D}_4$ in QIV represents the equiforce of the third component transaction—the decrease in the cash due to the payment of the current-period interest. The three equiforce vectors are not collinear so that they cannot be linearly added to get $\mathbf{D}$ as a single equiforce vector. Thus, the equiforce of the accounting event is represented by $\mathbf{D}_1$, $\mathbf{D}_3$, and $\mathbf{D}_4$ as a set.

If measured in the two-dimensional vector space, the power of the accounting event equals the size of the three DE vector spaces (squares) combined plus the size of the six rectangles combined. It is computed as

$$\begin{aligned} \mathrm{P} &= (|\mathbf{d}_1| + |\mathbf{d}_2| + |\mathbf{a}_1|)(|\mathbf{c}| + |\mathbf{a}_1| + |\mathbf{a}_2|) \\ &- (100 + 230 + |-120|)(100 + |-120| + |-230|) \\ &= 450^2 \mathrm{au}^2 \end{aligned}$$

The measure P of power above equals the space-based measure of the power of the accounting event seen in Sect. 5.3, Chap. 5. This event is 4.5 times larger in monetary size than the type 1 accounting event before. Hence, it is 20.25, not 4.5, times more powerful ($4.5^2 = 20.25$). This is because power is a multiplicative, not additive, concept and increases exponentially as the underlying force increases.

In the three-dimensional vector space, the accounting event has the following force vectors of order 3, all of which are coordinate vectors: $\mathbf{d}_1 = (100, 0, 0)$, $\mathbf{d}_2 = (230, 0, 0)$, $\mathbf{c} = (0, 100, 0)$, $\mathbf{b} = (-120, 0, 0)$, $\mathbf{a}_1 = (0, -120, 0)$, $\mathbf{a}_2 = (0, -230, 0)$, and $\mathbf{q}_1 = (0, 0, 100)$, $\mathbf{q}_2 = (0, 0, 120)$, and $\mathbf{q}_3 = (0, 0, 230)$. Some of these vectors, and the DE cube of the event, were shown in Fig. 6.10 before. As the accounting event has three component transactions, its equiforce vector is divided into the three discrete equiforce vectors: $\mathbf{D}_1$ in OI (not seen in the figure), $\mathbf{D}_3$ in OIII, and $\mathbf{D}_4$ in OIV. They are written as

$$\mathbf{D}_1 = \mathbf{d}_1 + \mathbf{c} + \mathbf{q}_1 = (100, 100, 100)$$

$$\mathbf{D}_3 = \mathbf{b} + \mathbf{a}_1 + \mathbf{q}_2 = (-120, -120, 120)$$

$$\mathbf{D}_4 = \mathbf{d}_2 + \mathbf{a}_2 + \mathbf{q}_3 = (230, -230, 230)$$

Since the three equiforce vectors above are not collinear, they cannot be linearly added to get **D** as a single equiforce vector.

The power of the accounting event is measured by the size of the large unshaded DE cube of the event. It is computed as

$$\begin{aligned} \mathrm{P} &= (|\mathbf{d}_1| + |\mathbf{d}_2| + |\mathbf{b}|)(|\mathbf{c}| + |\mathbf{a}_1| + |\mathbf{a}_2|)(|\mathbf{q}_1| + |\mathbf{q}_2| + |\mathbf{q}_3|) \\ &= (100 + 230 + |-120|)(100 + |-120| + |-230|)(450) \\ &= 450^3\,\mathrm{vu}^3 \end{aligned}$$

The space-based measure P of power above equals the combined size of the discrete DE cubes of the three component transactions plus the combined size of the 24 rectangular cuboids determined by the cross products of the debit, credit, and size forces of those component transactions.[8] It is equal to the space-based measure of the power of the accounting event in the three-dimensional space in Sect. 5.3, Chap. 5. When compared to the type 1 accounting event in the previous illustration, this accounting event is 4.5 times larger in monetary size and, hence, 91.125 times more powerful ($4.5^3 = 91.125$ or $20.25 \times 4.5 = 91.125$) as the power of the event is measured in the three-dimensional vector space. Again, this is because power is a multiplicative, not additive, concept and size matters when it comes to power. When measured as above, P is not one continuous magnitude but a multitude of discrete magnitudes. Measuring the power of the accounting event as such a multitude does not pose a problem since power in general is an absolute scalar quantity. Also, it is consistent with the notion that the "whole" is the same as the sum of its "parts."

6.6 Features of the Vectors and Vector Spaces of Accounting Events

The vectors and vector spaces of accounting events discussed thus far in this chapter have some interesting features. One feature concerns the notion of vector independence. In vector theory, vectors are considered mutually independent if they are orthogonal or point in different directions. As coordinate vectors, **d** and **c**, are orthogonal so that they are mutually independent in a mathematical sense. The same is also true for the coordinate vectors **d**, **c**, and **q**. In double entry accounting, however, they are *not* independent because, for an accounting event, they are interrelated, co-exist, and act together according to the double entry principle. In

[8] To see all this algebraically, the expression of P above can be expanded, which will give algebraic terms for the component DE cubes and rectangular cuboids, respectively.

this regard, they are dependent even if they point in different directions. For the same reason, **D** is also dependent although it points in a different direction because it is generated as a linear combination of **d** and **c**, or **d**, **c**, and **q**.

Another feature of the vectors of accounting events concerns what they mean conceptually. For a given accounting event, **d** and **c** as a pair or **d**, **c**, and **q** as a trio are coordinate vectors and are hence orthogonal. And they have the same magnitude under the double entry principle. As a set, they reflect the duality and balance features of the event and the contrasting yet symmetric nature of double entries. So do the DE vector **D** and DE vector space of the accounting event as they are defined by those coordinate vectors. Therefore, **D** has an angle of 45° with both coordinate vectors and a slope of 1 in the two-dimensional (debit-credit) vector space. And the DE vector space is a square. This square is not just any square but one of special kind (see Sect. 3.3, Chap. 3). It has two sides along, or in parallel with, the debit and credit coordinates. It also has a vertex at the point of origin of the two-dimensional vector space or the point of origin as an inner point. Any vector or square that does not have such features are not a DE vector or a DE vector space. It does not reflect the duality and balance features of accounting events. Nor does it reflect the contrasting yet symmetric nature of double entries.

What is said above also applies to the DE vectors and DE vector spaces of accounting events in the three-dimensional (debit-credit-size) vector space. In this space, the DE vector space of an accounting event is a cube. This cube is a DE cube, defined by **d**, **c**, and **q**, which have the same magnitude and act perpendicularly under the double entry principle. Conceptually, the DE cube reflects the duality and balance features of the accounting event and the contrasting yet symmetric nature of double entries. Thus, it is not just any cube but one of special kind (see Sect. 4.3, Chap. 4). It has a vertex at the point of origin of the vector space or the point of origin as an inner point. And it has two sides along, or are in parallel with, the debit and credit coordinates of the three-dimensional vector space. The diagonal of the cube is the DE vector **D** of the accounting event. **D** also acts according to the double entry principle and reflects the duality and balance features of the accounting event and the contrasting yet symmetric nature of double entries. It has an angle of 35.2645° with the bottom (debit-credit) plane of the three-dimensional vector space. If a cubic vector space does not have those features described above, it is not a DE vector space. Nor is the diagonal vector of such a vector space a DE vector.

Still another feature of the vectors of accounting events is about the magnitude of the DE vector **D**. This vector is a linear combination of **d** and **c**, for example, in the two-dimensional vector space. The three vectors **d**, **c**, and **D** define a right triangle as **d** and **c** are mutually orthogonal and have the same magnitude. Given the right triangle, the following hold according to the Pythagorean theorem:

$$
\begin{aligned}
|\mathbf{D}|^2 &= |\mathbf{d}|^2 + |\mathbf{c}|^2 \\
&= |\mathbf{d}|^2 + |\mathbf{c}|^2 + 2|\mathbf{d}||\mathbf{c}| \text{ cosine } \theta
\end{aligned}
$$

Thus,

$$|\mathbf{D}| = \left(|\mathbf{d}|^2 + |\mathbf{c}|^2\right)^{1/2}$$

where θ is the angle between **d** and **c**. Since θ = 90°, cosine θ = 0. The squared magnitude of **D** is equal to the sum of the squared magnitudes of **d** and **c**. That is, the magnitude of **D** is the same as the square root of the sum of the squared magnitudes of **d** and **c**. Since cosine θ = 0, 2|**d**||**c**| cosine θ = 0. Thus, the last expression above holds, which means that the Pythagorean theorem holds with respect to the vector space spanned by the three vectors. The squared magnitude of **D** above is related to what is called the dot product of vectors in vector theory, the dot product of **d** and **c** in this case, which is discussed in the next section.

6.7 Accounting Events in Vector Spaces: An Extension

In vector theory, there is a vector multiplication called the dot (inner or scalar) product of vectors. It is carried out by multiplying two vectors element by element. And the resultant products are then added up, which gives a scalar quantity (number). Thus, a dot product of vectors gives a scalar number—the reason why the vector multiplication is also called a scalar product of vectors. The dot product of two vectors (say, **u** and **v**) is written in general notations as

$$\mathbf{u}.\mathbf{v}^{\mathrm{T}} = (\mathrm{u}_1.\mathrm{v}_1) + (\mathrm{u}_2.\mathrm{v}_2) + \ldots + (\mathrm{u}_\mathrm{n}.\mathrm{v}_\mathrm{n}) = \mathbf{v}.\mathbf{u}^{\mathrm{T}}$$

where **u** and **v** are row vectors and $\mathbf{u}^{\mathrm{T}}$ and $\mathbf{v}^{\mathrm{T}}$ are their transpose (column vectors). The product is a scalar number. If, for example, **u** = (3, 0) and **v** = (3, 4), $\mathbf{u}.\mathbf{v}^{\mathrm{T}} = (3 \times 3 + 0 \times 4) = 9$, which is a scalar number. The vectors being multiplied should be conformable for vector multiplication, meaning that they should have the same order—the same number of elements. Also, the first (second) vector should be a row (column) vector.[9] A dot product of vectors in general is commutative. Thus, the dot product is the same, regardless of which vector is pre- or post-multiplied.

As a scalar number, the dot product of two vectors indicates how much the two vectors point in similar directions. If the scalar number is positive, this means that the two vectors point in similar directions and their interior angle is acute (0° < θ < 90°). If the dot product is negative, this means that the two vectors point in partly opposite directions and their interior angle is obtuse (90° < θ < 180°). The directional (angular) similarity between the two vectors is measured by cosine θ—the cosine of the angle between the two vectors. It is called their cosine similarity, and cosine θ is called their direction cosine. Given this, the dot product of **u** and **v** above can also be written as

[9] If the first (second) vector is a column (row) vector as well, the vector multiplication is a matrix multiplication of vectors, which is different from the dot product of vectors. It yields a matrix, not a scalar number.

$$\mathbf{u}.\mathbf{v}^{\mathrm{T}} = |\mathbf{u}||\mathbf{v}| \text{ cosine } \theta = \mathbf{v}.\mathbf{u}^{\mathrm{T}}$$

which means that

$$\text{cosine } \theta = \mathbf{u}.\mathbf{v}^{\mathrm{T}}/|\mathbf{u}||\mathbf{v}| \quad (|\mathbf{u}|, |\mathbf{v}| \neq 0)$$

If $\theta = 90°$, cosine $\theta = 0$ and the dot product is zero, meaning that **u** and **v** are orthogonal and have no directional similarity. They point in completely different directions. If $\theta = 0°$, cosine $\theta = 1$ and the dot product is not zero but equal to $|\mathbf{u}||\mathbf{v}|$, meaning that **u** and **v** are collinear and point in the same direction. If neither is the case ($0° < \theta < 90°$), cosine $\theta > 0$ but <1. And the angle between **u** and **v** is acute. The two vectors are not collinear, nor orthogonal, but point in similar directions.

Geometrically, the multiplication of **u** and **v** means that one of the two vectors is perpendicularly projected onto the other. This projection generates a new vector, which is called a projection vector. Specifically, $\mathbf{u}.\mathbf{v}^{\mathrm{T}}$ means that the second vector **v** is projected onto **u**, which generates a projection vector (say, $\mathbf{v}_{\mathrm{p}}$). This new vector is a perpendicular projection of **v** onto **u**. It is a null vector (point) if $\theta = 90°$ (cosine $\theta = 0$) and is colinear with **u** if $\theta = 0°$ (cosine $\theta = 1$). If the angle between **u** and **v** is acute ($0° < \theta < 90°$), $\mathbf{v}_{\mathrm{p}}$ is not a null vector, nor identical to **u**. Together with **v**, it defines a *right* triangle whose base is equal to $\mathbf{v}_{\mathrm{p}}$ and whose height is equal to $|\mathbf{v}—\mathbf{v}_{\mathrm{p}}|$, the distance between **v** and $\mathbf{v}_{\mathrm{p}}$. The side of the right triangle is equal to **v**.

While **v** and $\mathbf{v}_{\mathrm{p}}$ define the right triangle, the two vectors **u** and **v** span a parallelogram if their interior angle θ is acute. The shape and size of the parallelogram depend on the magnitude of the vectors and their interior angle. If $\theta = 90°$, cosine $\theta = 0$ and the two vectors are orthogonal. The dot product is zero, and $\mathbf{v}_{\mathrm{p}}$ is a null vector. And the parallelogram becomes a rectangle or square, depending on whether $|\mathbf{u}| = |\mathbf{v}|$ or not. If $\theta = 0°$, cosine $\theta = 1$ and the two vectors are collinear and point in the same direction. And the parallelogram degenerates into a null space (line). If neither is the case ($0° < \theta < 90°$), cosine $\theta > 0$ but <1. And the dot product is not zero, and the parallelogram is preserved.

Dot Products of the Vectors of Order 2 for Accounting Events The discussions above of the dot product of vectors in general and its geometric interpretations apply to the vectors of accounting events. Every accounting event has triple vectors of order 2 in the two-dimensional (debit-credit) vector space. They are $\mathbf{d} = (\mathrm{d}, 0)$, $\mathbf{c} = (0, \mathrm{c})$, and $\mathbf{D} = (\mathrm{d}, \mathrm{c})$. It is possible to have a dot product of each pair of these vectors. Given below are two examples.

$$\begin{aligned}\mathbf{d}.\mathbf{c}^{\mathrm{T}} &= (\mathrm{d}, 0)(0, \mathrm{c})^{\mathrm{T}} = \mathrm{d}.0 + 0.\mathrm{c} = 0\\ &= |\mathbf{d}||\mathbf{c}| \text{ cosine } \theta = 0\end{aligned}$$

$$\begin{aligned}\mathbf{c}.\mathbf{d}^{\mathrm{T}} &= (0, \mathrm{c})(\mathrm{d}, 0)^{\mathrm{T}} = 0.\mathrm{d} + \mathrm{c}.0 = 0\\ &= |\mathbf{c}||\mathbf{d}| \text{ cosine } \theta = 0\end{aligned}$$

which means

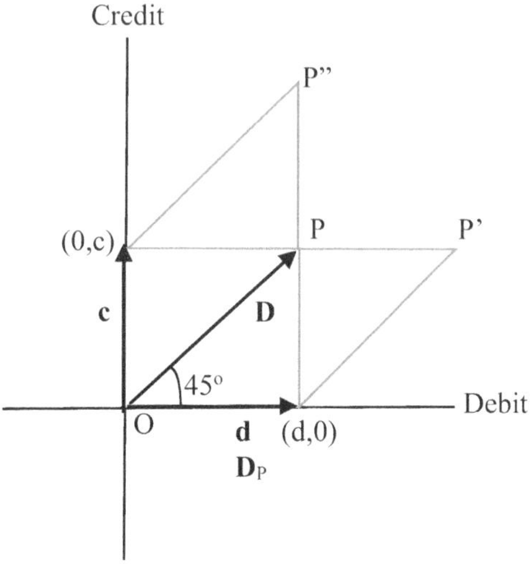

Fig. 6.11 Presenting the dot product of the vectors for type 1 accounting events in the two-dimensional vector space

$$\begin{aligned}\text{cosine }\theta &= \mathbf{d}.\mathbf{c}^T/|\mathbf{d}||\mathbf{c}| \text{ or } \mathbf{c}.\mathbf{d}^T/|\mathbf{c}||\mathbf{d}| \quad (|\mathbf{d}|, |\mathbf{c}| \neq 0)\\ &= 0\end{aligned}$$

where $\mathbf{d}^T$ and $\mathbf{c}^T$ are the transpose of **d** and **c** (row vectors) respectively and $\theta = 90°$, the angle between **d** and **c**. Since a dot product of vectors in general is commutative, $\mathbf{d}.\mathbf{c}^T = \mathbf{c}.\mathbf{d}^T$. Each vector multiplication above yields a scalar number, which is zero because **d** and **c** are orthogonal and cosine $90° = 0$. If, for example, a firm sells goods for \$100 in cash, $\mathbf{d}.\mathbf{c}^T = (100, 0)(0, 100)^T = 0$ or $\mathbf{c}.\mathbf{d}^T = (0, 100)(100, 0)^T = 0$. This means that **d** and **c** are orthogonal so that they have no directional similarity. Since $\mathbf{d}.\mathbf{c}^T = 0$ and $|\mathbf{d}||\mathbf{c}| > 0$, $\mathbf{d}.\mathbf{c}^T < |\mathbf{d}||\mathbf{c}|$. Thus, the well-known Cauchy–Schwarz inequality in vector theory holds with respect to **d** and **c**. It says that the magnitude of the dot product of two vectors cannot be greater than the product of their magnitudes.[10] The geometric space spanned by **d** and **c** is not a parallelogram but a square—a DE square, since **d** and **c** are orthogonal and $|\mathbf{d}| = |\mathbf{c}|$ under the double entry principle.

The geometric meanings of the dot product $\mathbf{d}.\mathbf{c}^T$ for an accounting event can be examined graphically. Figure 6.11 shows what the dot product $\mathbf{d}.\mathbf{c}^T$ (or $\mathbf{c}.\mathbf{d}^T$) means in the two-dimensional (debit-credit) vector space. The accounting event in the figure is assumed to be of type 1 and increase the debit and credit accounts alike. Thus, both the DE vector space (unshaded square) and the DE vector **D** lie in the first quadrant (QI) of the vector space. The three arrows shown are the debit, credit, and DE vectors respectively of the accounting event. As with a dot product of two vectors in general, $\mathbf{d}.\mathbf{c}^T$ indicates how much **d** and **c** are similar in their directional orientation. That is, it tells how much of **c** points in the direction of **d**, or vice versa.

[10] If the two vectors are **u** and **v**, for example, the Cauchy–Schwarz inequality holds as $|\mathbf{u}.\mathbf{v}| \leq |\mathbf{u}||\mathbf{v}|$. For details, see a standard textbook of linear algebra or an *Internet* website.

Since **d** and **c** are orthogonal ($\theta = 90°$), cosine $\theta = 0$ and $\mathbf{d}.\mathbf{c}^T = 0$. This means that the two vectors have no directional similarity. They are orthogonal and point in completely different directions.

The dot product $\mathbf{d}.\mathbf{c}^T$ also means that the vector multiplication perpendicularly projects **c** onto **d**, or vice versa. This projection generates a new vector—a projection vector (say, $\mathbf{c}_P$, not shown in the figure). This new vector is an orthogonal projection of **c** onto **d**. It is a null vector ($\mathbf{c}_p = \mathbf{0}$) and $|\mathbf{c}_P| = 0$ since **d** and **c** are orthogonal ($\theta = 90°$). It is a point—the point of origin. The triangle defined by **c** and $\mathbf{c}_P$ degenerates into a null space—the point of origin. As **d** and **c** are orthogonal and $|\mathbf{d}| = |\mathbf{c}|$ under the double entry principle, the vector space spanned by **d** and **c** is a square, not a parallelogram. This square is the DE square (OdPc) shown in the figure.

The dot product $\mathbf{c}.\mathbf{d}^T$ is interpreted similarly. Since the two vectors are orthogonal ($\theta = 90°$), $\mathbf{c}.\mathbf{d}^T = 0$, meaning that the two vectors have no directional similarity. The vector multiplication perpendicularly projects **d** onto **c**, which generates a projection vector (say, $\mathbf{d}_p$, not shown in the figure). Since **c** and **d** are orthogonal and have the same magnitude under the double entry principle, the projection vector is a null vector ($\mathbf{d}_p = \mathbf{0}$), and its magnitude is zero ($|\mathbf{d}_P| = 0$). It is a point—the point of origin. The triangle defined by **d** and $\mathbf{d}_P$ degenerates into a null space—the point of origin. And the vector space spanned by **c** and **d** is a square—the same DE square shown in the figure. The two dot products $\mathbf{d}.\mathbf{c}^T$ and $\mathbf{c}.\mathbf{d}^T$ each span the same DE square. This is consistent with the fact that the dot product of vectors in general is commutative. It is also consistent with the duality and balance features of accounting events or, more generally, the duality and balance features of double entry accounting.

Besides the dot products above, there are other dot products that can be defined with the vectors of an accounting event. They are

$$\begin{aligned}\mathbf{d}.\mathbf{D}^T &= (\mathrm{d}, 0)(\mathrm{d}, \mathrm{c})^T = \mathrm{d.d} + 0.\mathrm{c} = \mathrm{d}^2 \\ &= |\mathbf{d}||\mathbf{D}|\text{cosine}\,\theta = |\mathbf{d}|^2\end{aligned}$$

$$\begin{aligned}\mathbf{c}.\mathbf{D}^T &= (0, \mathrm{c})(\mathrm{d}, \mathrm{c})^T = 0.\mathrm{d} + \mathrm{c.c} = \mathrm{c}^2 \\ &= |\mathbf{c}||\mathbf{D}|\text{cosine}\,\theta = |\mathbf{c}|^2\end{aligned}$$

which means

$$\begin{aligned}\text{cosine}\,\theta &= \mathbf{d}.\mathbf{D}^T/|\mathbf{d}||\mathbf{D}| \quad (|\mathbf{d}|, |\mathbf{D}| \neq 0) \\ &= \mathbf{c}.\mathbf{D}^T/|\mathbf{c}||\mathbf{D}| \quad (|\mathbf{c}|, |\mathbf{D}| \neq 0) \\ &\approx 0.7071\end{aligned}$$

where **D** is the DE vector of the accounting event and $\theta = 45°$. And **D** cosine $\theta =$ **d** or **c** according to the law of cosine. Unlike the case of $\mathbf{d}.\mathbf{c}^T$ and $\mathbf{c}.\mathbf{d}^T$ before, each dot product above is not zero since the two vectors being multiplied are not orthogonal. If, for example, a firm sells goods for \$100 in cash, $\mathbf{d} = (100, 0)$, $\mathbf{c} = (0, 100)$, $\mathbf{D} = (100, 100)$, and $|\mathbf{D}| = 141.4214$. Thus, $\mathbf{d}.\mathbf{D}^T = (100, 0)(100, 100)^T = 100^2$, which is equal to d^2. Or $\mathbf{d}.\mathbf{D}^T = (100)(141.4214)(0.70711) = 100^2$. Since the dot product of vectors in general is commutative, $\mathbf{d}.\mathbf{D}^T = \mathbf{D}.\mathbf{d}^T$ and $\mathbf{c}.\mathbf{D}^T = \mathbf{D}.\mathbf{c}^T$. As the angle between **d** and **D** is acute, cosine $\theta \neq 0$. Thus, $\mathbf{d}.\mathbf{D}^T = |\mathbf{d}|^2$ or d^2 (or d.c), which is a nonzero scalar quantity (number). As $-1 \leq$ cosine $\theta \leq +1$ in general, $\mathbf{d}.\mathbf{D}^T \leq |\mathbf{d}||\mathbf{D}|$ and $|\mathbf{d}.\mathbf{D}^T| / |\mathbf{d}||\mathbf{D}| \leq 1$. Thus, the Cauchy–Schwarz inequality holds with respect to **d** and **D**. As the angle between **d** and **D** is acute, $0 <$ cosine $\theta < +1$, meaning that **d** and **D** are neither collinear nor orthogonal. They point not in the same direction, nor in opposite directions, but in similar directions.

As with $\mathbf{d}.\mathbf{c}^T$ or $\mathbf{c}.\mathbf{d}^T$ before, the dot products $\mathbf{d}.\mathbf{D}^T$ and $\mathbf{c}.\mathbf{D}^T$ have similar geometric interpretations. As for $\mathbf{d}.\mathbf{D}^T$, the vector multiplication perpendicularly projects **D** onto **d** (or vice versa), which generates a projection vector. This new vector is shown by $\mathbf{D}_P$ in Fig. 6.11. It is a shadow of **D** projected onto **d**. Since $\theta = 45°$, $\mathbf{D}_P = \mathbf{d}$ and $|\mathbf{D}_P| = |\mathbf{d}| = |\mathbf{D}|$ cosine θ. The two vectors **D** and $\mathbf{D}_P$ define the right triangle OdP, whose base is equal to $\mathbf{D}_P$ and whose height is equal to $|\mathbf{D} - \mathbf{D}_P|$.[11] Given $\mathbf{D}_P$, the directional similarity between **d** and **D** is shown by the direction cosine:

$$\begin{aligned} \text{cosine}\,\theta &= \mathbf{d}.\mathbf{D}^T/|\mathbf{d}||\mathbf{D}| = |\mathbf{d}|/|\mathbf{D}| = |\mathbf{D}_P|/|\mathbf{D}| \\ &\approx 0.7071 \end{aligned}$$

While **d** and $\mathbf{D}_P$ cannot be divided by **D** as vectors in general have a direction, their magnitude can be divided by the magnitude of **D**. Given this, the directional similarity between **d** and **D** as measured by cosine θ is equal to 0.7071, which is equivalent to the angle of 45°. And **D** has a slope of 1 as tangent 45° = 1 (sine 45° = cosine 45°). The direction cosine is positive but less than 1 so that **d** and **D** point in similar directions, not in the same direction or in perpendicular directions.

As **d** and **D** have an acute angle, the vector space spanned by them is not a square or rectangle but a parallelogram—the parallelogram OPP'd in the figure. The size of this parallelogram is equal to $|\mathbf{d}||\mathbf{D}_P|$ or $|\mathbf{d}|^2$, which is equal to the size of the DE vector space (square) defined by **d** and **c**. The base of the parallelogram is equal to $|\mathbf{d}|$ and its height is equal to $|\mathbf{c}|$. The size of the parallelogram is equal to $|\mathbf{d}||\mathbf{c}|$ or just d.c, namely the size of the DE square. For a given accounting event, therefore, the following hold in absolute terms under the double entry principle:

$$\begin{aligned} |\mathbf{d}.\mathbf{D}^T| &= |\mathbf{d}||\mathbf{D}|\ \text{cosine}\,\theta \\ &= |\mathbf{d}||\mathbf{d}| = |\mathbf{d}|^2 = |\mathbf{d}||\mathbf{c}| = |\text{d.c}| \end{aligned}$$

where **D** cosine $\theta = \mathbf{d}$. What the above means is that the magnitude of $\mathbf{d}.\mathbf{D}^T$ is equal to the size of the DE square of the accounting event and that the size of the parallelogram is equal to the size of the DE square of the event:

$$\text{OPP}'\text{d} = \text{OdPc}$$

[11] The vector subtraction $\mathbf{D} - \mathbf{D}_P$ may be seen as a difference vector, and $|\mathbf{D} - \mathbf{D}_P|$ as its magnitude.

The dot product $\mathbf{c}.\mathbf{D}^T$ is interpreted similarly. It means that the vector multiplication perpendicularly projects **D** onto **c**, which generates a projection vector ($\mathbf{D}_P$, not shown in the figure). Since θ = 45°, $\mathbf{D}_P = \mathbf{c}$ and $|\mathbf{D}_P| = |\mathbf{c}| = |\mathbf{D}|$ cosine θ. The dot product $\mathbf{c}.\mathbf{D}^T$ shows a directional similarity between **c** and **D**, which is measured by cosine $\theta = \mathbf{c}.\mathbf{D}^T / |\mathbf{c}||\mathbf{D}| \approx 0.7071$. That is, θ = 45°, meaning that the two vectors are partly similar in direction. The vector space spanned by **c** and **D** is the parallelogram OPP"c in the figure. This parallelogram is symmetric to the parallelogram OPP'd spanned by **d** and **D**. And it has the same size as the DE square: OPP"c = OdPc.

Given the dot products and their geometric meanings above, and under the double entry principle, the following hold in absolute terms:

$$\begin{gathered} |\mathbf{d}||\mathbf{c}| = \left|\mathbf{d}.\mathbf{D}^T\right| = \left|\mathbf{c}.\mathbf{D}^T\right| \\ \mathrm{OdPc} = \mathrm{OPP'd} = \mathrm{OPP''c} \end{gathered}$$

That is, the three vector spaces—the DE square and two symmetric parallelograms—in Fig. 6.11 are the same in size. This is so because **d**, **c**, and **D** for an accounting event are interrelated and act together according to the double entry principle and **D** reflects the duality and balance features of the accounting event. The two parallelograms are contrasting yet symmetric to each other and have the same size, which is consistent with the duality and balance features of the accounting event or, more generally, the contrasting yet symmetric nature of the double entry principle. The same is also true for the two right triangles in each parallelogram. The size of each triangle, whether spanned by **D** and **d** or **D** and **c**, is exactly one half of the size of the DE vector space (square) defined by **d** and **c** or **c** and **d** and exactly one half of the size of the parallelogram defined by $\mathbf{d}.\mathbf{D}^T$ and $\mathbf{c}.\mathbf{D}^T$, respectively.

In addition to the various dot products above, there are other dot products of the vectors of accounting events. They are the dot product of **d**, **c**, and **D** respectively with *itself* as shown below.

$$\begin{aligned} \mathbf{d}.\mathbf{d}^T &= (\mathrm{d}, 0)(\mathrm{d}, 0)^T = \mathrm{d}.\mathrm{d} + 0.0 = \mathrm{d}^2 \\ &= |\mathbf{d}||\mathbf{d}|\text{cosine } \theta = |\mathbf{d}|^2 \end{aligned}$$

$$\begin{aligned} \mathbf{c}.\mathbf{c}^T &= (0, \mathrm{c})(0, \mathrm{c})^T = 0.0 + \mathrm{c}.\mathrm{c} = \mathrm{c}^2 \\ &= |\mathbf{c}||\mathbf{c}|\text{cosine } \theta = |\mathbf{c}|^2 \end{aligned}$$

$$\begin{aligned} \mathbf{D}.\mathbf{D}^T &= (\mathrm{d}, \mathrm{c})(\mathrm{d}, \mathrm{c})^T = \mathrm{d}^2 + \mathrm{c}^2 = 2\mathrm{d}^2 \text{ or } 2\mathrm{c}^2 \\ &= |\mathbf{D}||\mathbf{D}|\text{cosine } \theta = |\mathbf{D}|^2 \end{aligned}$$

And

$$\begin{aligned} \text{cosine } \theta &= \mathbf{d}.\mathbf{d}^T/|\mathbf{d}||\mathbf{d}| = 1 \quad (|\mathbf{d}| \neq 0) \\ &= \mathbf{c}.\mathbf{c}^T/|\mathbf{c}||\mathbf{c}| = 1 \quad (|\mathbf{c}| \neq 0) \\ &= \mathbf{D}.\mathbf{D}^T/|\mathbf{D}||\mathbf{D}| = 1 \quad (|\mathbf{D} \neq 0) \end{aligned}$$

where $\theta = 0°$ and cosine $\theta = 1$. As seen, each dot product above yields a scalar quantity (number), and cosine θ is equal to 1. The square root of each dot product equals the magnitude of the vector itself. Also, each dot product indicates a directional similarity between the vector and itself. Since $\theta = 0°$, cosine $\theta = 1$, meaning that the two vectors being multiplied are identical—the same vector—and point in the same direction. And the parallelogram defined by the vectors degenerates into a null space (line).

Given the dot products above, and the right triangles defined by **d**, **c**, and **D**, the following hold for an accounting event:

$$\mathbf{D}.\mathbf{D}^{T} = \mathbf{d}.\mathbf{d}^{T} + \mathbf{c}.\mathbf{c}^{T}$$
$$|\mathbf{D}|^{2} = |\mathbf{d}|^{2} + |\mathbf{c}|^{2}$$

That is, $\mathbf{D}.\mathbf{D}^{T}$ is a linear addition of $\mathbf{d}.\mathbf{d}^{T}$and $\mathbf{c}.\mathbf{c}^{T}$, which is consistent with the definition of **D**—that the DE vector of accounting events is a linear combination of **d** and **c**: $\mathbf{D} = \mathbf{d} + \mathbf{c}$. Similarly, $|\mathbf{D}|^{2}$ is a linear addition of $|\mathbf{d}|^{2}$ and $|\mathbf{c}|^{2}$. This means that the Pythagorean theorem holds with respect to the three vectors, which jointly span the right triangle OdP (or OcP) in Fig. 6.11. The two expressions above are equivalent. Conceptually, they reflect the duality and balance features of accounting events or, more generally, the contrasting yet symmetric nature of the double entry principle.

The dot products and their interpretations thus far concern type 1 accounting events that increase the single debit and credit accounts alike. If an accounting event is of type 1 and increases the debit account but decreases the credit account, the dot products can be written as $\mathbf{d}.\mathbf{a}^{T}$ and $\mathbf{a}.\mathbf{D}^{T}$, where **a** is a vector representing the monetary quantity of the credit account as a decrease. The expressions and interpretations of those dot products are similar to those of the various dot products shown before.

The dot products of the vectors of accounting events of type 2, 3, or 4 can be written similarly and interpreted in the same way. A difference is that the dot products of the vectors of such accounting events are defined in terms of the discrete vectors relating to their component transactions. For such accounting events, there are multiple sets of dot products, one for each component transaction. For the kth component transaction of such accounting events, for example, the dot product of $\mathbf{d}_k$ and $\mathbf{D}_k$ is written as $\mathbf{d}_k.D_k{}^{T} = |\mathbf{d}_k||\mathbf{D}_k|$ cosine θ. Yet the features and interpretations of those dot products are similar to those of the dot products of the vectors of type 1 accounting events described before.

Dot Products of the Vectors of Order 3 for Accounting Events The discussions thus far of the dot products of the vectors of order 2 for accounting events can be extended to the dot products of those vectors of order 3 in the three-dimensional (debit-credit-size) vector space. In this space, there is another coordinate vector—the size vector **q**. Thus, for every accounting event, there are four different vectors: **d**, **c**,

q, and **D**. Given this, the various dot products of those vectors can be written as shown below.

$$\begin{aligned}\mathbf{d}.\mathbf{c}^{T} &= (d,0,0)(0,c,0)^{T} = d.0 + 0.c + 0.0 = 0 = \mathbf{c}.\mathbf{d}^{T}\\ &= |\mathbf{d}||\mathbf{c}| \text{ cosine } \theta = 0\end{aligned}$$

$$\begin{aligned}\mathbf{d}.\mathbf{q}^{T} &= (d,0,0)(0,0,q)^{T} = d.0 + 0.0 + 0.q = 0 = \mathbf{q}.\mathbf{d}^{T}\\ &= |\mathbf{d}||\mathbf{q}| \text{ cosine } \theta = 0\end{aligned}$$

$$\begin{aligned}\mathbf{c}.\mathbf{q}^{T} &= (0,c,0)(0,0,q)^{T} = 0.0 + c.0 + 0.q = 0 = \mathbf{q}.\mathbf{c}^{T}\\ &= |\mathbf{c}||\mathbf{q}| \text{ cosine } \theta = 0\end{aligned}$$

$$\begin{aligned}\mathbf{d}.\mathbf{D}^{T} &= (d,0,0)(d,c,q)^{T} = d.d + 0.c + 0.q = d^2 = \mathbf{D}.\mathbf{d}^{T}\\ &= |\mathbf{d}||\mathbf{D}| \text{ cosine}\,\theta \text{ or } |\mathbf{c}||\mathbf{D}| \text{ cosine}\,\theta\end{aligned}$$

$$\begin{aligned}\mathbf{c}.\mathbf{D}^{T} &= (0,c,0)(d,c,q)^{T} = 0.d + c.c + 0.q = c^2 = \mathbf{D}.\mathbf{c}^{T}\\ &= |\mathbf{c}||\mathbf{D}| \text{ cosine}\,\theta \text{ or } |\mathbf{d}||\mathbf{D}| \text{ cosine}\,\theta\end{aligned}$$

$$\begin{aligned}\mathbf{q}.\mathbf{D}^{T} &= (0,0,q)(d,c,q)^{T} = 0.d + 0.c + q.q = q^2 = \mathbf{D}.\mathbf{q}^{T}\\ &= |\mathbf{q}||\mathbf{D}| \text{ cosine}\,\theta\end{aligned}$$

The interpretations of each dot product above are similar to those of the dot products in the two-dimensional space before, except that the geometric setting is three-dimensional and the vectors are all of order 3. Each dot product yields a scalar number, positive or negative, and indicates a directional similarity between the two vectors being multiplied. Also, the vectors in each dot product span a vector space. For example, **d** and **q** in the dot product $\mathbf{d}.\mathbf{q}^{T}$ span a square in the three-dimensional vector space, and **d** and **D** in the dot product $\mathbf{d}.\mathbf{D}^{T}$ span a parallelogram in the three-dimensional vector space.

As with the two-dimensional case before, the dot product of each vector with itself is also possible in the three-dimensional setting. They are

$$\begin{aligned}\mathbf{d}.\mathbf{d}^{T} &= (d,0,0)(d,0,0)^{T} = d^2\\ &= |\mathbf{d}||\mathbf{d}| \text{ cosine } \theta = |\mathbf{d}|^2\end{aligned}$$

$$\begin{aligned}\mathbf{c}.\mathbf{c}^{T} &= (0,c,0)(0,c,0)^{T} = c^2\\ &= |\mathbf{c}||\mathbf{c}| \text{ cosine } \theta = |\mathbf{c}|^2\end{aligned}$$

$$\begin{aligned}\mathbf{q}.\mathbf{q}^{T} &= (0,0,q)(0,0,q)^{T} = q^2\\ &= |\mathbf{q}||\mathbf{q}| \text{ cosine } \theta = |\mathbf{q}|^2\end{aligned}$$

$$\mathbf{D}.\mathbf{D}^{\mathrm{T}} = (\mathrm{d}, \mathrm{c}, \mathrm{q})(\mathrm{d}, \mathrm{c}, \mathrm{q})^{\mathrm{T}} = \mathrm{d}^2 + \mathrm{c}^2 + \mathrm{q}^2 = 3\mathrm{d}^2, 3\mathrm{c}^2, \text{or } 3\mathrm{q}^2$$
$$= |\mathbf{D}||\mathbf{D}| \text{ cosine } \theta = |\mathbf{D}|^2$$

As seen, the dot product of each vector above is equal to the squared magnitude of the vector itself.

Given the various dot products above, the following also hold, as in the two-dimensional case before:

$$\mathbf{D}.\mathbf{D}^{\mathrm{T}} = \mathbf{d}.\mathbf{d}^{\mathrm{T}} + \mathbf{c}.\mathbf{c}^{\mathrm{T}} + \mathbf{q}.\mathbf{q}^{\mathrm{T}}$$
$$|\mathbf{D}|^2 = |\mathbf{d}|^2 + |\mathbf{c}|^2 + |\mathbf{q}|^2$$

As seen, $\mathbf{D}.\mathbf{D}^{\mathrm{T}}$ is a linear addition of $\mathbf{d}.\mathbf{d}^{\mathrm{T}}$, $\mathbf{c}.\mathbf{c}^{\mathrm{T}}$, and $\mathbf{q}.\mathbf{q}^{\mathrm{T}}$. This is consistent with the definition of **D** as a DE vector: $\mathbf{D} = \mathbf{d} + \mathbf{c} + \mathbf{q}$. And $|\mathbf{D}|^2$ is a linear addition of $|\mathbf{d}|^2$, $|\mathbf{c}|^2$, and $|\mathbf{q}|^2$. Thus, the Pythagorean theorem holds with respect to those vectors in the three-dimensional vector space. Conceptually, each expression above reflects the duality and balance features of accounting events or, more generally, the contrasting yet symmetric nature of the double entry principle.

The various dot products of the vectors of order 3 shown before concern accounting events of type 1. If an accounting event is of type 2, 3, or 4, having component transactions, each of the vectors of the event is divided into multiple discrete vectors. Thus, there are multiple sets of dot products of those discrete vectors. Each of these dot products can be written in similar ways the various dot products of the vectors of type 1 accounting events were before. And their features and interpretations are similar to those of the dot products of the vectors of type 1 accounting events discussed before.

The Cross Product of the Vectors of Accounting Events

Besides the dot product of vectors, there is another kind of vector multiplication—the cross product of vectors. This differs from a dot product of vectors in several respects. Unlike a dot product of vectors, it is not a row-by-column, or column-by-row, multiplication. Instead, the multiplication of vectors is performed according to the well-known cofactor expansion rule in matrix algebra, which is used to compute the determinant of matrices (see Sect. 7.2, Chap. 7). Also, unlike a dot product of vectors, a cross product of vectors is defined only in a three-dimensional vector space so that the vectors being cross multiplied are of order 3. And a cross product of vectors yields a vector, not a scalar number.

Shown below is a cross product of two vectors of order 3 in general notations.

$$\begin{aligned}\mathbf{uxv} &= (u_1, u_2, u_3) \times (v_1, v_2, v_3) \neq \mathbf{vxu} \\ &= ((+1)(u_2v_3 - u_3v_2), (-1)(u_1v_3 - u_3v_1), (+1)(u_1v_2 - u_2v_1)) \\ &= (u_2v_3 - u_3v_2, u_3v_1 - u_1v_3, u_1v_2 - u_2v_1) \\ &= (w_1, w_2, w_3) = \mathbf{w}\end{aligned}$$

Each element of the new vector **w** is the difference between the cross products of the elements of **u** and **v**, and the sign (+1 or −1) of each element of **w** is determined according to the cofactor expansion rule.[12] The cross product of vectors is distributive but, unlike a dot product, is not commutative. Hence, the resultant vector **w** differs, depending on which vector is pre- or post-multiplied. If, for example, **u** = (3, 1, 1) and **v** = (3, 2, 1), **uxv** = ((+1)(1 × 1 − 2 × 1), (−1)(3 × 1 − 3 × 1), (+1) (3 × 2 − 3 × 1)) = (−1, 0, 3) = **w**, which is not the same as **vxu** = (1, 0, −3).

Considering the angle between **u** and **v**, the cross product of the two vectors can also be written as

$$\mathbf{uxv} = |\mathbf{u}||\mathbf{v}| \text{ sine } \theta \mathbf{e} = \mathbf{w}$$

where θ is the angle between **u** and **v** and **e** is a three-element unit vector. If a vector has a magnitude of 1, it is called a unit vector, regardless of its order.[13] And |**e**| = 1. The expression of **uxv** is different from the expression of the dot product $\mathbf{u}.\mathbf{v}^T$ before. It includes sine θ, versus cosine θ, and the unit vector **e**.

Unlike the dot product $\mathbf{u}.\mathbf{v}^T$ before, the cross product **uxv** yields a vector (**w**), not a scalar number. The magnitude of **uxv**, namely the magnitude |**w**|, indicates how much the two vectors are dissimilar in direction. This directional dissimilarity is shown by sine θ—the sine of the angle between **u** and **v**. As sine changes inversely with cosine, so is the cross product of vectors inversely related to the dot product of the same vectors. In other words, the directional dissimilarity of **u** and **v** is inversely related to their directional similarity. As **u** and **v** become more dissimilar in direction, they become less similar. And vice versa. If the two vectors are collinear (θ = 0°), sine θ = 0. Hence, **uxv** = **w** = **0**, meaning that their cross product yields a null vector. And the magnitude |**w**| decreases to its minimum, which is zero. What this means is that the two vectors are collinear and have no dissimilarity in direction. If the two vectors are orthogonal (θ = 90°), sine θ = 1. Their cross product yields a vector, which is not a null vector: **uxv** = **w** ≠ **0**. The two vectors are completely dissimilar in direction. They point in perpendicular directions. And the magnitude of |**w**| increases to its maximum.

[12] The second element of the expression is multiplied by −1, while the first and third elements are multiplied by +1. This is due to the way the cofactor expansion rule works. About this rule, see Sect. 7.2, Chap. 7.

[13] For example, the vectors (1, 0), (0, 1, 0), and (1/√2, 1/√2), and (1/√3, 1/√3, 1/√3) are all unit vectors as their magnitude is 1. The elements of a unit vector do not need to be all 1. In fact, vectors with all elements equal to 1, such as (1, 1) and (1, 1, 1), are not a unit vector since their magnitude is not 1.

Geometrically, the cross product of two vectors means that the vector multiplication projects the resultant vector **w** straight up if $0° < \theta < 180°$, or down if $180° < \theta < 360°$, from the plane (flat area) where the two vectors being cross multiplied lie in the three-dimensional space. The magnitude of **w** varies with the angle between the two vectors. The larger the angle, the greater is the magnitude |**w**|. So is their directional dissimilarity. And vice versa. The magnitude of **w** increases to its maximum if **u** and **v** are orthogonal ($\theta = 90°$ or $270°$) and decreases to its minimum (zero) if the two vectors are collinear or point in opposite directions ($\theta = 0°$ or $180°$). In this case, **w** is a null vector. Also, the two vectors being cross multiplied span a parallelogram if their interior angle is acute. And the magnitude of **w** is equal to the size of this parallelogram.

The discussions thus far of the cross product of vectors in general apply to the cross product of the vectors of accounting events in the three-dimensional (debit-credit-size) vector space. In this vector space, accounting events have four vectors: **d** = (d, 0, 0), **c** = (0, c, 0), **q** = (0, 0, q), and **D** = (d, c, q). Given these vectors, it is possible to write their cross products in various forms. Following the cofactor expansion rule mentioned before, the cross product of **d** and **c** can be written as

$$\begin{aligned}\mathbf{dxc} &= (\mathrm{d}, 0, 0) \times (0, \mathrm{c}, 0)\\ &= ((+1)(0.0 - \mathrm{c}.0), (-1)(\mathrm{d}.0 - 0.0), (+1)(\mathrm{d.c} - 0.0))\\ &= (0, 0, \mathrm{d.c}) = \mathbf{w}\end{aligned}$$

where **w** is a new vector of order 3. If, for example, a firm sells goods for \$100 in cash, **d** = (100, 0, 0) and **c** = (0, 100, 0). Thus, **dxc** = $(0, 0, 100^2)$ = **w**. Since a cross product of vectors in general is *not* commutative, **dxc** ≠ **cxd**. For the transaction example, **cxd** = $(0, 0, -100^2)$ = **w**, which is not the same as **w** above.

The magnitude of the cross product, namely |**w**|, depends on the angle between **d** and **c**. Given this, **dxc** can also be written as

$$\mathbf{dxc} = |\mathbf{d}||\mathbf{c}|\text{sine}\ \theta\,\mathbf{e} = |\mathbf{d}||\mathbf{c}|\mathbf{e} = \mathbf{w}$$

Since **d** and **c** are orthogonal,

$$\begin{aligned}&|\mathbf{dxc}|/|\mathbf{d}||\mathbf{c}| = 1 \quad (|\mathbf{d}|, |\mathbf{c}| \neq 0)\\ &|\mathbf{dxc}| = |\mathbf{d}||\mathbf{c}|\\ &\text{sine}\ \theta = 1\end{aligned}$$

where $\theta = 90°$, sine $90° = 1$, and **e** is a unit vector of order 3 whose magnitude is 1 (|**e**| = 1). The magnitude of **dxc** is equal to the product of the magnitudes of the two vectors being cross multiplied: |**dxc**| = |**d**||**c**|.

The expression of the cross product **dxc** can be compared to the expression of the dot product $\mathbf{d.c}^{\mathrm{T}}$ before. Besides the vectors being of order 3 versus order 2, **dxc** differs from $\mathbf{d.c}^{\mathrm{T}}$, not only in the multiplication of the elements of the vectors but also in the presence (absence) of **e** and sine θ versus cosine θ. Each element of the new vector **w** is the difference between the cross products of the elements of **d** and **c**, and its sign (+1 or −1) is determined according to the cofactor expansion rule.

The cross product **dxc** has interesting geometric meanings. Unlike the dot product $\mathbf{d}.\mathbf{c}^T$, it yields a vector (**w**), which points straight up from the debit-credit plane of the three-dimensional (debit-credit-size) vector space. As such, the new vector **w** is perpendicular to **d** and **c**. Also, the cross product **dxc** indicates a directional dissimilarity between **d** and **c**. The angular dissimilarity between **d** and **c** is equivalent to their angular similarity since the two are inversely related to each other. As sine θ changes inversely with cosine θ, the directional dissimilarity between **d** and **c** measured by sine θ is inversely related to the directional similarity between them measured by cosine θ. The magnitude |**dxc**|, namely |**w**|, increases to its maximum as **d** and **c** are orthogonal (θ = 90°) and sine θ = 1, meaning that **d** and **c** are completely dissimilar in directional orientation.

The cross product **dxc** also means that **d** and **c** span a vector space. This space is not a parallelogram but a square (area) as the two vectors are orthogonal and |**d**| = |**c**| under the double entry principle. This square is the DE square of the accounting event in question on the debit-credit plane of the three-dimensional vector space. As sine θ = 1 (θ = 90°) and |**e**| = 1, |**dxc**| = |**d**||**c**|. That is, the magnitude of **dxc**, namely the magnitude |**w**|, is equal to the size of the DE square spanned by **d** and **c**. If, for example, a firm sells goods for $100, **d** = (100, 0, 0) and **c** = (0, 100, 0). And |**dxc**| = |**d**||**c**| sine θ |**e**| = (100)(100)(1)(1) = 100^2, which is equal to the size of the DE square of the accounting transaction represented by **d** and **c**. The size of this square can also be computed as |**d**||**c**| = d × c = 100 × 100 = 100^2.

For a given accounting event, it is also possible to define other cross products of the debit, credit, and size coordinate vectors as shown below.

$$\mathbf{cxd} = |\mathbf{c}||\mathbf{d}| \text{ sine } \theta\, \mathbf{e} = |\mathbf{c}||\mathbf{d}|\mathbf{e} = \mathbf{w}$$
$$\mathbf{dxq} = |\mathbf{d}||\mathbf{q}| \text{ sine } \theta\, \mathbf{e} = |\mathbf{d}||\mathbf{q}|\mathbf{e} = \mathbf{w}$$
$$\mathbf{cxq} = |\mathbf{c}||\mathbf{q}| \text{ sine } \theta\, \mathbf{e} = |\mathbf{c}||\mathbf{q}|\mathbf{e} = \mathbf{w}$$

where θ = 90° and sine θ = 1 and **w** is the new vector in each case. The multiplication of the vectors in each cross product above is performed similarly as **dxc** was before. And their interpretations are similar. Also, each cross product of the vectors is not commutative, so **cxd** ≠ **dxc**, for example. The pair of the coordinate vectors being cross multiplied is mutually orthogonal (θ = 90°) so that sine θ =1. They are orthogonal and, hence, completely dissimilar in direction. The magnitude of the cross product, namely |**w**|, increases to its maximum as the two vectors being cross multiplied are orthogonal. Also, it is equal to the size of the vector space spanned by the two vectors. This vector space is a square under the double entry principle. In the case of **cxd** in particular, it is the DE square of the accounting event and the magnitude |**w**| is equal to the size of the DE square.

Besides the coordinate vectors, every accounting event has another vector—the DE vector **D**. Given this vector, the following cross products can also be defined:

$$\begin{aligned}\mathbf{dxD} &= (\mathrm{d},0,0)\times(\mathrm{d},\mathrm{c},\mathrm{q}) = (0,(-1)\mathrm{d.q},\mathrm{d.c}) = (0,-\mathrm{d.q},\mathrm{d.c}) = \mathbf{w}\\ &= |\mathbf{d}||\mathbf{D}|\ \mathrm{sine}\,\theta\,\mathbf{e} = \mathbf{w}\end{aligned}$$

$$\begin{aligned}\mathbf{cxD} &= (0,\mathrm{c},0)(\mathrm{d},\mathrm{c},\mathrm{q}) = (\mathrm{c.q},(-1)(0),-\mathrm{d.c}) = (\mathrm{c.q},0,-\mathrm{d.c}) = \mathbf{w}\\ &= |\mathbf{c}||\mathbf{D}|\ \mathrm{sine}\,\theta\,\mathbf{e} = \mathbf{w}\end{aligned}$$

$$\begin{aligned}\mathbf{qxD} &= (0,0,\mathrm{q})(\mathrm{d},\mathrm{c},\mathrm{q}) = (-\mathrm{c.q},(-1)(-\mathrm{d.q}),0) = (-\mathrm{c.q},\mathrm{d.q},0) = \mathbf{w}\\ &= |\mathbf{q}||\mathbf{D}|\ \mathrm{sine}\,\theta\,\mathbf{e} = \mathbf{w}\end{aligned}$$

where θ is the angle between the two vectors being cross multiplied ($\theta \approx 54.735°$) and **w** is the new vector in each case.[14] **D** has the same angle with **d**, **c**, and **q** respectively as the three coordinate vectors are orthogonal and have the same magnitude. Since a cross product of vectors in general is not commutative, **dxD** ≠ **Dxd**, **cxD** ≠ **Dxc**, and **qxD** ≠ **Dxq**.

The multiplication of the vectors in each cross product above is interpreted in similar ways the cross product **dxc** was before. An exception is that, since the two vectors being cross product are not orthogonal, the vector space spanned by the vectors is a parallelogram, not a square or rectangle, with its sides equal to the two vectors being cross multiplied. And the magnitude of **dxD**, namely |**w**|, is equal to the size of the parallelogram spanned by **d** and **D** in the three-dimensional vector space. If, for example, a firm sells goods for \$100, **d** = (100, 0, 0), **D** = (100, 100, 100), the magnitude of **D** and **dxD** respectively is calculated as follows:

$$\begin{aligned}|\mathbf{D}| &= \left(100^2+100^2+100^2\right)^{1/2} = 173.2051\\ |\mathbf{dxD}| &= |\mathbf{d}||\mathbf{D}|\mathrm{sine}\ \theta\ |\mathbf{e}| = (100)(173.2051)(0.816497)(1)\\ &= 14,142.14\end{aligned}$$

The magnitude |**dxD**| is equal to the size of the parallelogram spanned by **d** and **D**. Algebraically, the size of this parallelogram is computed as: base × height = 100 × 141.4214 = 14,142.14. The base of the parallelogram is 100, which is equal to |**d**|, and the height is 141.4214, which is equal to |**D**| sine θ: 173.2051 × sine 54.735°.

[14] This angle can be found as follows: if a firm sells goods for \$100, for example, |**d**| = |**c**| = |**q**| = 100. And $|\mathbf{D}| = (|\mathbf{d}|^2 + |\mathbf{c}|^2 + |\mathbf{q}|^2)^{1/2} = 173.2051$ according to the Pythagorean theorem. Cosine θ = |**d**| / |**D**| = 100 / 173.2051 = 0.57735 or sine $\theta = (|\mathbf{c}|^2 + |\mathbf{q}|^2)^{1/2}$ / |**D**| = 141.42136 / 173.2051 = 0.81650. Thus, θ ≈ 54.735°.

The Double Entry Accounting and Matrix Algebra

7

In matrix algebra, a matrix is defined as an array of real numbers. An accounting event (transaction) has three intrinsic elements—debit, credit, and size, denoted by d, c, and q in previous chapters. As the monetary quantity of each element is shown by a real number, accounting events can be shown as an array of real numbers, hence as a matrix. This chapter discusses how to construct the matrix of accounting events and examines its features and properties from the perspective of matrices and theories thereof and according to the double entry principle. The geometric settings used to do so are two- and three-dimensional vector spaces with multiple Cartesian coordinates. There are various kinds of matrices, and the chapter focuses on two of them in particular—diagonal matrix and scalar matrix. These two kinds of matrices have features and properties that are consistent with the duality and balance features of accounting events or, more generally, the duality and balance features of double entry accounting.

7.1 The Matrix of Accounting Events

Matrices and Related Concepts

A matrix is a square or rectangular array of real numbers. Each number is called an element of the matrix. If the elements of a matrix are arranged as a single row (or column), the matrix is a row (column) vector. If they are arranged as multiple rows (columns), the matrix is a square or rectangular matrix. It is a square matrix if it has the same number of rows and columns. Otherwise, it is a rectangular matrix. The number of the rows and columns of a matrix is called the size, order, or dimension of the matrix. If a matrix has m rows and n columns, it is a matrix of size $m \times n$ (or order $m \times n$) or just an $m \times n$ matrix.

Related to the order of matrices is something called the rank of matrices. The rank of a matrix is the maximum number of its linearly independent rows (or columns). Not all the rows or columns of a matrix may be mutually independent. Thus, the rank

© The Author(s), under exclusive license to Springer Nature Switzerland AG 2025

B. T. Ro, *The Geometry of Accounting*, Springer Texts in Business and Economics,
https://doi.org/10.1007/978-3-031-88892-2_7

of a matrix is not always the same as the order of the matrix. The rank of an m x n matrix cannot be greater than the smaller of m or n: $r(\mathbf{A}) \leq \min(m,n)$, where $r(\mathbf{A})$ is the rank of **A**. The procedure of finding the rank of a matrix can be simple or complex, depending on the size of the matrix. The rank of a matrix is found by performing the so-called elementary row or column operation in matrix algebra. This is a mathematical operation that reduces the matrix in row- or column-echelon form. Every matrix can be reduced in row- or column-echelon form. The rank of the matrix is equal to the number of the pivot rows (or columns) of the matrix in is reduced echelon form.[1]

Besides rectangular and square matrices defined above, there are other kinds of matrices. If each row of a square matrix is the same as each column, the matrix is a symmetric matrix. That is, **A** is a symmetric matrix if $\mathbf{A} = \mathbf{A}^T$, where $\mathbf{A}^T$ is a transpose of **A**. All symmetric matrices are a square matrix, but the converse is not true. By definition, a rectangular matrix cannot be a symmetric matrix. One special kind of symmetric matrix is a diagonal matrix. This is a square matrix whose off-diagonal elements are all zero. If the main diagonal elements are zero as well, the diagonal matrix is a null matrix. If the main diagonal elements of a diagonal matrix are the same (nonzero) scalar number, the matrix is called a scalar matrix. Thus, all scalar matrices are diagonal matrices, but the converse is not true.

A scalar matrix is a special kind of diagonal matrix in that its main diagonal elements are the same (nonzero) scalar number, positive or negative. As such, it can be written as a nonzero scalar number multiplied by a unit (identity) matrix. That is, if **A** is a scalar matrix, it can be written as $\mathbf{A} = k\mathbf{I}$, where k is a (nonzero) scalar number and **I** is a unit matrix. A unit matrix, also called an identity matrix, is a diagonal matrix whose main diagonal elements are all unity (1).

Geometrically, each column of a matrix represents a point in a vector space, and each row represents a dimension (coordinate) of the vector space. Or vice versa. Only the interpretations differ. If a matrix has multiple columns, and not all of these columns are dependent, the matrix represents a set of noncollinear points, which define a space. This space is a vector space represented by the matrix. A vector space represented by a 2×2 matrix is a flat area (e.g., rectangle or parallelogram), and a vector space represented by a 3×3 matrix is a cubic space (e.g., cube or rectangular cuboid). This is consistent with the definition of a geometric space in general, which is defined by a set of noncollinear points, three of them at a minimum. Similarly, the vector space represented by a matrix is defined by a set of noncollinear column (or row) vectors of the matrix, each representing a distinct point. The independent columns (rows) of a matrix are considered a set of basis vectors and, as a set, they span the vector space represented by the matrix. This vector space is also called the

[1]The first nonzero element of each row (column) of a matrix in reduced echelon form is called a pivot of the matrix. The row (column) that has a pivot is called a pivot row (column). The pivot element is then changed to a unity (1) by dividing the row (column) by the pivot element. The rank of the matrix is equal to the number of such pivot rows (columns). If every resultant element of the row (column) is zero, the row (column) has no pivot, and this row (column) is not counted in determining the rank of the matrix.

column (row) space of the matrix. And it is the same as the vector space represented by the matrix itself.

The Double Entry Matrix of Accounting Events

An accounting event has three intrinsic elements—debit, credit, and size, which were denoted by d, c, and q in quantity form in Sect. 2.1, Chap. 2. Each of those elements is shown by a real number. Geometrically, it is considered a point on the respective coordinate of the vector space where the accounting event is presented. Together, those points define a geometric space in the multidimensional vector space. They are shown by real numbers, which can be arranged as a matrix. And this matrix represents the accounting event. How it can be constructed depends on the dimensionality of the vector space where the accounting event is presented.

The Double Entry Matrix in Two-Dimensional Setting In the two-dimensional vector space with two Cartesian (debit and credit) coordinates, the matrix (say, **M**) representing an accounting event is a 2 × 2 diagonal or scalar matrix. Shown below is how this matrix can be constructed according to the double entry principle in the two-dimensional (debit-credit) vector space.

$$[\text{Debit}\quad \text{Credit}] \rightarrow \mathbf{M} = \begin{bmatrix} \mathbf{d} \\ \mathbf{c} \end{bmatrix} = \begin{bmatrix} \mathrm{d} & 0 \\ 0 & \mathrm{c} \end{bmatrix}$$

$$\begin{matrix} & & \text{Dimension (coordinate)} & \\ & & \text{Debit}\quad\text{Credit} & \\ & & \downarrow \qquad \downarrow & \\ \begin{bmatrix} \text{Debit} & \text{xx} \\ \text{Credit} & \text{xx} \end{bmatrix} & \rightarrow \mathbf{M} = & \begin{bmatrix} \mathrm{d} & 0 \\ 0 & \mathrm{c} \end{bmatrix} & \begin{matrix} \leftarrow \text{Debit (point d} > \text{or} < 0) \\ \leftarrow \text{Credit (point c} > \text{or} < 0) \end{matrix} \end{matrix}$$

The way in which **M** is constructed above is not consistent with a convention in matrix algebra—that each column of a matrix represents a point and each row represents a dimension (coordinate) of the two-dimensional (debit-credit) vector space. Instead, **M** is constructed by considering its columns representing the debit and credit dimensions of the two-dimensional vector space and its rows representing the points of the debit and credit entries on those dimensions. Specifically, the first column of **M** represents the debit dimension of accounting events, and the second column represents the credit dimension. And the first row of **M** represents the point for the debit element of accounting events, and the second row represents the point for the credit element. Since there are only two dimensions in the two-dimensional vector space, the size dimension of accounting events is suppressed. As it is, **M** is constructed according to not the convention in matrix algebra but the journal-entry format [debit credit] or the layout of the debit and credit columns of a journal book in accounting as a basic frame. Hereafter, **M** is called the double entry (DE) matrix of accounting events.

The DE matrix **M** has a pair of row vectors: **d** = (d, 0) and **c** = (0, c). The vector **d** represents the point (d,0) for the debit element d of accounting events on the debit coordinate, and **c** represents the point (0,c) for the credit element c on the credit coordinate. The two row vectors have the same magnitude and point in perpendicular directions under the double entry principle. The debit element d means an increase or a decrease in the debit account, depending on the nature of the account (e.g., asset or liability). And the credit element c means an increase or a decrease in the credit account, depending on the nature of the account (e.g., liability or asset). Thus, the main diagonal elements d and c of **M** can be a positive or negative number (d, c > or < 0).

For an accounting event, **M** is a diagonal or scalar matrix, depending on whether or not d and c have the same sign (+ or −) as an increase or a decrease in monetary quantity. Regardless, |d| = |c| under the double entry principle. **M** is a 2 × 2 symmetric matrix and, conceptually, considered to reflect the duality and balance features of the accounting event and the contrasting yet symmetric nature of double entries. **M** can also be a scalar matrix if the main diagonal elements d and c are the same scalar number, positive or negative (+ or −). That is, it is a scalar matrix if the accounting event changes the debit and credit accounts in the same direction, increase or decrease. Regardless, **M** is always a symmetric matrix so that interchanging its rows and columns does not change its features and properties; only the interpretations differ. By nature, **M** cannot be a null matrix since no accounting event can take place with a zero monetary volume (d, c ≠ 0). Nor can it be a rectangular matrix since the double entry principle does not allow it to be.

As a 2 × 2 diagonal or scalar matrix, **M** has a rank of 2. If it is a 2 × 2 scalar matrix in particular, the main diagonal elements are the same scalar number, positive or negative. What this means is that the accounting event represented by **M** either increases or decreases the debit and credit accounts alike and do so to the full extent of its monetary size. As a scalar matrix, **M** can be written as: **M** = d**I** or **M** = c**I,** where d and c are the same scalar number and **I** is a 2 × 2 unit matrix. This unit matrix is considered to reflect the duality and balance features of the accounting event in matrix form.

As a diagonal or scalar matrix, **M** has two row vectors, **d** and **c**, which are mutually orthogonal and have the same magnitude (|**d**| = |**c**|) under the double entry principle. Thus, the vector space represented by **M** is a square, not a rectangle or parallelogram. It is defined by the pair of orthogonal row vectors (points) that have the same magnitude in the two-dimensional vector space. It is the row space of the matrix as it is spanned by the row vectors of **M**. This space is represented by **M**. It is a double entry (DE) vector space, equivalent to the DE vector space of accounting events in Chap. 6 and the DE square in Chap. 3. It reflects the same features of **M**—the duality and balance features of accounting events and the contrasting and symmetric nature of double entries. It is spanned by **d** and **c** as a pair of basis vectors.

Figure 7.1 gives an example of a DE vector space represented by **M**. The shaded square shown is the DE vector space, defined by four vertices (points): (0,0), (d,0), (0,c), and (d,c). Each vertex can be determined by multiplying **M** with the following column vectors: (0, 0), (1, 0), (0, 1), and (1, 1) respectively, which jointly define a

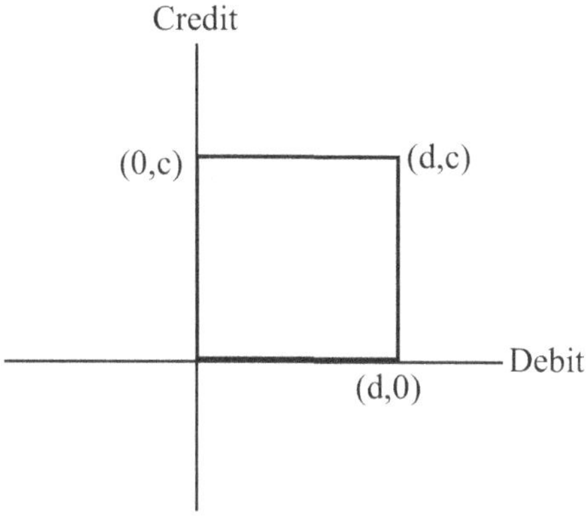

Fig. 7.1 A vector space represented by the double entry (DE) matrix **M** of type 1 accounting events in the two-dimensional vector space

vector space (square) whose size is 1. The shaded DE vector space has the same features as **M**. It reflects the duality and balance features of the accounting event and the contrasting yet symmetric nature of double entries.

The DE vector space in the figure concerns an accounting event of type 1, having no component transactions and increasing the single account on both debit and credit sides. Hence, it lies in the first quadrant (QI) of the debit-credit vector space. If an accounting event is of the same type and increases the debit account but decreases the credit account, the diagonal elements of **M** are not the same scalar number (+ or −) although their magnitude is the same. Thus, **M** is a diagonal, not scalar, matrix. And the DE vector space represented by **M** lies in QIV, although it is still a square. When all such possibilities are considered, **M** for type 1 accounting events be written in one of the following four forms, depending on how the accounting events affect the debit and credit accounts:

$$\mathbf{M} = \begin{bmatrix} \mathbf{d} \\ \mathbf{c} \end{bmatrix} = \begin{bmatrix} d & 0 \\ 0 & c \end{bmatrix} \text{or} \begin{bmatrix} b & 0 \\ 0 & c \end{bmatrix} \text{or} \begin{bmatrix} b & 0 \\ 0 & a \end{bmatrix} \text{or} \begin{bmatrix} d & 0 \\ 0 & a \end{bmatrix}$$

where d, c > 0 (increase in monetary quantity) and b, a < 0 (decrease in monetary quantity).[2] The DE matrices shown above are all diagonal matrices and, hence, symmetric matrices.

The DE matrix **M** in the first and third forms above is a scalar matrix as their main diagonal elements are the same scalar number. Thus, it can be written as $\mathbf{M} = d\mathbf{I}$ or $c\mathbf{I}$ or $\mathbf{M} = b\mathbf{I}$ or $a\mathbf{I}$, where **I** is a 2×2 unit matrix. **M** in the second and fourth forms is a diagonal matrix since their main diagonal elements are not the same scalar number. Regardless, **M** in each form above is a diagonal matrix, hence a symmetric matrix, and reflects the duality and balance features of the accounting events. The DE vector space (square) represented by the first DE matrix above concerns type 1 accounting events that increase the debit and credit accounts alike. Hence, it lies in QI of the

[2]The debit and credit quantity notations d, c, b, and a are consistent with the notations used to denote the cell blocks of the tableau of the double entry system in Fig. 2.3, Chap. 2 earlier.

debit-credit vector space. The DE vector space (square) represented by the third matrix concerns type 1 accounting events that decrease the debit and credit accounts alike. Hence, it lies in QIII. The second DE matrix concerns type 1 accounting events with the debit account as a decrease and the credit account as an increase. The DE vector space represented by this matrix lies in QII. The fourth DE matrix is the opposite of the second one, so the DE vector space lies in QIV.

If an accounting event is of type 2, 3, or 4, having multiple entries (accounts) on the debit or credit side (or both sides), it is considered a bundle of component transactions. As such, each diagonal element of **M** is represented by multiple numbers, positive or negative. Such an accounting event is represented by not a single DE matrix but a set of DE matrices—one for each component transaction. The DE matrix of each component transaction has one of the following four forms as **M** for type 1 accounting events does:

$$\mathbf{M}_1 = \begin{bmatrix} d_k & 0 \\ 0 & c_k \end{bmatrix} \mathbf{M}_2 = \begin{bmatrix} b_k & 0 \\ 0 & c_k \end{bmatrix} \mathbf{M}_3 = \begin{bmatrix} b_k & 0 \\ 0 & a_k \end{bmatrix} \text{ or } \mathbf{M}_4 = \begin{bmatrix} d_k & 0 \\ 0 & a_k \end{bmatrix}$$

where the diagonal elements of each matrix above are the debit and credit of the kth component transaction of the accounting event. The subscripts 1, 2, 3, and 4 denote the quadrants of the debit-credit vector space where the DE vector space of each component transaction lies. The DE matrices above are all diagonal matrices and, hence, symmetric matrices. And $\mathbf{M}_1$ and $\mathbf{M}_3$ in particular are scalar matrices as their main diagonal elements are the same scalar number, positive or negative: $d_k = c_k$ and $b_k = a_k$.

The Double Entry Matrix in Three-Dimensional Setting Much of the discussions above of how to construct **M** and interpret it in the two-dimensional case thus far applies to the three-dimensional (debit-credit-size) case, except that there is one more dimension—the (monetary) size dimension of accounting events. As such, **M** is a 3 × 3 diagonal or scalar matrix and has a rank of 3. As in the two-dimensional case before, it is always a symmetric matrix. If an accounting event is of type 1, **M** has one of the following four forms, which are equivalent to their counterparts of **M** in the two-dimensional case before:

$$\mathbf{M} = \begin{bmatrix} \mathbf{d} \\ \mathbf{c} \\ \mathbf{q} \end{bmatrix} = \begin{bmatrix} d & 0 & 0 \\ 0 & c & 0 \\ 0 & 0 & q \end{bmatrix} \begin{bmatrix} b & 0 & 0 \\ 0 & c & 0 \\ 0 & 0 & q \end{bmatrix} \begin{bmatrix} b & 0 & 0 \\ 0 & a & 0 \\ 0 & 0 & q \end{bmatrix} \text{ or } \begin{bmatrix} d & 0 & 0 \\ 0 & a & 0 \\ 0 & 0 & q \end{bmatrix}$$

The vectors **d**, **c**, and **q** represent d, c, and q respectively of the accounting event. As coordinate vectors, they are mutually orthogonal and have the same magnitude: $|\mathbf{d}| = |\mathbf{c}| = |\mathbf{q}|$ under the double entry principle. So do the main diagonal elements: $|d| = |c| = |q|$,..., and $|d| = |a| = |q|$. If they have the same sign (+ or −) as well, **M** is a 3 × 3 scalar matrix, such as the first matrix above. Otherwise, it is a 3 × 3 diagonal matrix. Regardless, **M** in all forms is a symmetric matrix. As a scalar matrix, the first

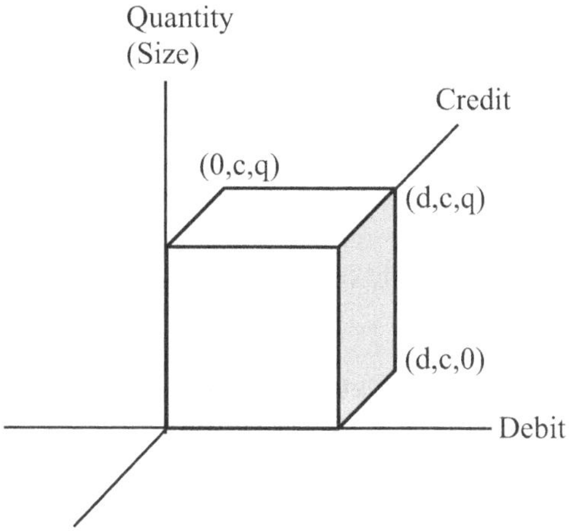

Fig. 7.2 A vector space represented by the double entry (DE) matrix M of type 1 accounting events in the three-dimensional vector space

matrix can be written as **M** = d**I**, c**I**, or q**I,** where **I** is a 3 × 3 unit matrix and reflects the duality and balance features of the accounting event.

Figure 7.2 gives an example of the DE vector space represented by **M** in the three-dimensional setting. The accounting event in the figure is assumed to be of type 1 and increase the debit and credit accounts alike (d, c > 0). Thus, **M** is a 3 × 3 scalar matrix. And the shaded vector space shown is the DE cube of the accounting event represented by **M** and lies in the first octant (OI) of the three-dimensional vector space. If an accounting event decreases the debit and credit accounts alike, **M** is the same as the third matrix above. The DE cube represented by this matrix lies in OIII. The second DE matrix represents an accounting event with the debit account as a decrease and the credit account as an increase. Thus, the DE cube represented by this matrix lies in OII. Opposite to the second matrix is the fourth matrix. The DE cube of this accounting event lies in OIV.

If an accounting event is of type 2, 3, or 4, it has component transactions, which are executed in bundle as the event is executed. Thus, the debit, credit, and size elements respectively of the accounting event are shown by multiple numbers. And the event is represented by not a single DE matrix but a set of discrete DE matrices, one for each component transaction. These discrete matrices are all 3 × 3 diagonal matrices and have a rank of 3. And the DE matrix of each component transaction has one of the following four forms, similar to those of **M** for type 1 accounting events before:

$$\mathbf{M}_1 = \begin{bmatrix} d_k & 0 & 0 \\ 0 & c_k & 0 \\ 0 & 0 & q_k \end{bmatrix} \mathbf{M}_2 = \begin{bmatrix} b_k & 0 & 0 \\ 0 & c_k & 0 \\ 0 & 0 & q_k \end{bmatrix} \mathbf{M}_3 = \begin{bmatrix} b_k & 0 & 0 \\ 0 & a_k & 0 \\ 0 & 0 & q_k \end{bmatrix}$$

$$\mathbf{M}_4 = \begin{bmatrix} d_k & 0 & 0 \\ 0 & a_k & 0 \\ 0 & 0 & q_k \end{bmatrix}$$

where d_k (or b_k), c_k (or a_k), and q_k are the debit, credit, and size elements respectively of the kth component transaction of the accounting event. The subscripts 1, 2, 3, and 4 denote the octant of the three-dimensional vector space in which the DE cube of the component transaction represented by the related DE matrix lies. $\mathbf{M}_1$ is a 3 × 3 scalar matrix, and the other three matrices are all a diagonal matrix. Representing a component transaction, each DE matrix above reflects the duality and balance features of the related component transaction and the contrasting yet symmetric nature of double entries.

7.2 The Determinant of the Double Entry Matrix

The Determinant of Matrices

Matrices have something called a determinant, which is a positive or negative scalar number. Thus, the determinant of matrices is often shown in absolute value bars. It is a multiplicative, not an additive, concept as its computation involves the cross multiplication of the elements of matrices. Square matrices have a determinant, but not all square matrices do so; only those that are nonsingular have a (nonzero) determinant. A square matrix is nonsingular if it has an inverse, meaning that it has a nonzero determinant. Otherwise, it is singular and has no inverse. While not all square matrices have a determinant, all diagonal matrices have a nonzero determinant and, hence, are nonsingular. The same is also true for scalar matrices since scalar matrices are part of the diagonal matrix family.

The computation of the determinant of a matrix may be simple or quite involved, depending on how large the size (dimension) of the matrix is. The simplest is the case of a 2 × 2 square matrix, as shown below.

$$\begin{vmatrix} a & b \\ c & d \end{vmatrix} = |ad - cb|$$

The computation of the determinant of matrices of higher order is not as simple. One common way to compute the determinant of such matrices is to use the well-known Laplace expansion formula, also called the cofactor expansion rule, in matrix algebra. Given below is an example of how to compute the determinant of a 3 × 3 matrix (say, $\mathbf{A}$) under this rule.

$$|\mathbf{A}| = \begin{vmatrix} a_{11} & a_{12} & a_{13} \\ a_{21} & a_{22} & a_{23} \\ a_{31} & a_{32} & a_{33} \end{vmatrix} = a_{11}(+1)^2 \begin{vmatrix} a_{22} & a_{23} \\ a_{32} & a_{33} \end{vmatrix} + a_{12}(-1)^3 \begin{vmatrix} a_{21} & a_{23} \\ a_{31} & a_{33} \end{vmatrix}$$

$$+a_{13}(+1)^2 \begin{vmatrix} a_{21} & a_{22} \\ a_{31} & a_{32} \end{vmatrix}$$

$$= a_{11}(+1)M_{11} + a_{12}(-1)M_{12} + a_{13}(+1)M_{13}$$

$$= a_{11}C_{11} + a_{12}C_{12} + a_{13}C_{13}$$

$$= a_{11}|a_{22}.a_{33} - a_{32}.a_{23}| - a_{12}|a_{21}.a_{33} - a_{31}.a_{23}| + a_{13}|a_{21}.a_{32} - a_{31}.a_{22}|$$

In this example, the first row of **A** is chosen to compute |**A**| under the cofactor expansion rule, and a_{11}, a_{12}, and a_{13} are the chosen row elements to begin the process of computation under the rule. The submatrix, minor, and cofactor for a_{11}, a_{12}, and a_{13} respectively are then generated. A submatrix for each chosen row element is formed with the rows and columns of the matrix *not* related to the chosen row element. The determinant of the submatrix is then computed. It is called a minor. This gives the three minors M_{11}, M_{12}, and M_{13} for the three chosen row elements above. Each minor has a positive or negative sign, and a signed minor is called a cofactor.[3] The determinant of **A** is obtained by adding the products of the chosen row elements and the related cofactors. As shown, the computation of the determinant of a matrix is heavily involved in the cross multiplication of its elements—the reason why the determinant of matrices is a multiplicative, not an additive, concept.

The determinant of matrices has interesting properties. The determinant of a matrix is zero if its rows or columns are collinear or dependent. It is symmetric: $|\mathbf{A}| = |\mathbf{A}^T|$. That is, interchanging the rows and columns of **A** does not affect its determinant. Also, the determinant of the inverse of a matrix is equal to the inverse of its determinant: $|\mathbf{A}^{-1}| = |\mathbf{A}|^{-1}$. However, the determinant of the sum of two matrices is not the same as the sum of their determinants: $|\mathbf{A} + \mathbf{B}| \neq |\mathbf{A}| + |\mathbf{B}|$. This is because the determinant of matrices is a multiplicative, not an additive, concept.

When viewed geometrically, the determinant of a matrix in absolute value is equal to the size of the vector space represented by the matrix. This vector space is a flat area in a two-dimensional vector space and a cubic space in a three-dimensional vector space. Also, the determinant of a matrix indicates how large the vector space represented by the matrix is relative to the vector space represented by a unit matrix. The determinant of a unit matrix is 1, and the size of the vector space represented by a unit matrix is also 1. Given this, the determinant of a matrix is equal to the ratio of the size of the vector space represented by the matrix to the size of the vector space represented by a unit matrix. If, for example, the determinant of a matrix is 10, this means that the vector space represented by the matrix is 10 times larger than the vector space represented by a unit matrix.

The Determinant of the Double Entry Matrix

As a 2 × 2 or 3 × 3 diagonal matrix, the DE matrix of accounting events has a determinant. The computation of this determinant, as well as its magnitude and interpretations, depends on the types of accounting events and the dimensionality of the vector space where accounting events are presented.

[3] The sign of each minor is determined by the factor $(-1)^{i+j}$, where i and j are the ith row and jth column of a_{ij}, the chosen row element. This factor is +1 if "i + j" is an even number and − 1 if it is an odd number. In the example above, there are three minors and, hence, three cofactors in the example: $C_{11} = (+1)M_{11}$, $C_{12} = (-1)M_{12}$, and $C_{13} = (+1)M_{13}$.

The Two-Dimensional Case In the two-dimensional (debit-credit) vector space, the DE matrix **M** of an accounting event is a 2×2 diagonal matrix. And the determinant of **M** is computed as

$$|\mathbf{M}| = \begin{vmatrix} \mathrm{d} & 0 \\ 0 & \mathrm{c} \end{vmatrix} = |\mathrm{d.c} - 0.0| = |\mathrm{d.c}|$$

The determinant $|\mathbf{M}|$ is equal to $|\mathbf{d}||\mathbf{c}|$ or just $|\mathrm{d.c.}|$. As **M** is a symmetric matrix, $|\mathbf{M}| = |\mathbf{M}^{\mathrm{T}}|$. That is, the determinant of **M** is the same even if the rows and columns of **M** are interchanged, which is consistent with the contrasting yet symmetric nature of the double entry principle.

As with the determinant of matrices in general, the determinant of **M** has similar geometric interpretations. It is equal in magnitude to the size of the vector space represented by **M**, which is equal to $|\mathrm{d.c.}|$. This vector space is a DE square, such as the one in Fig. 7.1. It is a square since **d** and **c** are orthogonal and $|\mathbf{d}| = |\mathbf{c}|$, namely $\mathrm{d} = \mathrm{c}$, under the double entry principle. The location of the DE square in the two-dimensional debit-credit vector space depends on the sign (+ or −) of the main diagonal elements d and c of **M**. It lies in the first or third quadrant (QI or QIII) of the two-dimensional vector space if d and c have the same sign (+ or −); otherwise, it lies in QII or QIV. The accounting event represented by the shaded DE square in the figure is assumed to be of type 1 and increase the debit and credit accounts alike. Thus, the DE square lies in QI.

If an accounting event is of type 2, 3, or 4, thus having component transactions, it is represented by a set of discrete DE matrices—$\mathbf{M}_1$ and $\mathbf{M}_2$ or $\mathbf{M}_1$ and $\mathbf{M}_4$, for example. The diagonal elements of these discrete matrices may or may not have the same sign (+ or −). Each discrete matrix has a distinct determinant of its own. However, the determinant of **M** for such accounting events cannot be computed as a simple sum of the determinants of those discrete DE matrices. This is because of the general property of determinants: $\mathbf{A} + \mathbf{B}| \neq |\mathbf{A}| + |\mathbf{B}|$, for example. Again, this is because the determinant of matrices is a multiplicative, not an additive, concept.

Given the above, one way to compute the determinant of the DE matrix of type 2, 3, or 4 accounting events is to, first, construct a special DE matrix (say, $\mathbf{M}_{\mathrm{S}}$) with the diagonal elements of **M** in absolute terms. The determinant of $\mathbf{M}_{\mathrm{S}}$ is then computed as usual. Computing the determinant of the DE matrix of such accounting events this way is reasonable, given that the determinant of matrices in general is a directionless scalar quantity.

Shown below are the special DE matrix of type 2, 3, or 4 accounting events in general notations and the computation of its determinant.

$$\mathbf{M}_S = \begin{bmatrix} \sum|d_k| & 0 \\ 0 & \sum|c_k| \end{bmatrix}$$

$$|\mathbf{M}_S| = \begin{bmatrix} \sum|d_k| & 0 \\ 0 & \sum|c_k| \end{bmatrix} = \left|\sum|d_k|\sum|c_k| - 0.0\right| = \sum|d_k|\sum|c_k|$$

The determinant $|\mathbf{M}_S|$ equals the size of the DE vector space represented by $\mathbf{M}_S$. This vector space can lie in any quadrant of the two-dimensional (debit-credit) vector space, or across multiple (but not three) quadrants, depending on the sign (+ or −) of d_k and c_k.

Alternatively, the determinant of $\mathbf{M}_S$ for a type 2, 3, or 4 accounting event is computed by, first, computing the determinant of the DE matrix of each component transaction and the cross products of the main diagonal elements of the DE matrices of the component transactions. And those determinants and cross products are then added, all in absolute terms, which gives a scalar quantity. This scalar quantity is considered equal to the determinant of $\mathbf{M}_S$. If, for example, an accounting event has two component transactions and is represented by $\mathbf{M}_1$ and $\mathbf{M}_2$, the determinant of $\mathbf{M}_S$ is computed as

$$\mathbf{M}_1 = \begin{bmatrix} d_1 & 0 \\ 0 & c_1 \end{bmatrix} \quad \mathbf{M}_2 = \begin{bmatrix} d_2 & 0 \\ 0 & c_2 \end{bmatrix}$$

$$\begin{aligned} |\mathbf{M}_S| &= |\mathbf{M}_1| + |\mathbf{M}_2| + |\mathbf{d}_1||\mathbf{c}_2| + |\mathbf{d}_2||\mathbf{c}_1| \\ &= |d_1.c_1| + |d_2.c_2| + |d_1.c_2| + |d_2.c_1| \end{aligned}$$

$|\mathbf{M}_1|$ and $|\mathbf{M}_2|$ are the determinant of the DE matrix of the respective component transactions. $|\mathbf{d}_1||\mathbf{c}_2|$ and $|\mathbf{d}_2||\mathbf{c}_1|$, or $|d_1.c_2|$ and $|d_2.c_1|$, respectively are the cross product of the debit (credit) of one component transaction and the credit (debit) of the other. The direct product $|d_1.c_1|$ is equal to the size of the DE square represented by $\mathbf{M}_1$, and $|d_2.c_2|$ is equal to the size of the DE square represented by $\mathbf{M}_2$. And the cross product $|d_1.c_2|$ is equal to the size of the rectangle (or square) defined by $d_1$and c_2. Similarly, $|d_2.c_1|$ is equal to the size of the rectangle (or square) defined by d_2 and c_1. The size of the DE square of the accounting event equals the size of the two DE squares combined plus the size of the two rectangles combined. Computing $|\mathbf{M}_S|$ as described above is reasonable, given that the determinant of matrices in general is a directionless scalar quantity as indicated before. It is consistent with the "whole-parts" notion—that the "whole" is the same as the sum of its "parts." The DE square of the accounting event is considered the "whole" and the component squares and rectangles are considered the "parts."

The Three-Dimensional Case Much of the discussions above of the determinants of DE matrices in the two-dimensional case applies to the three-dimensional (debit-credit-size) case, except that $\mathbf{M}$ is now a 3×3 diagonal matrix. For a given accounting event, $\mathbf{M}$ has the three row vectors $\mathbf{d}$, $\mathbf{c}$, and $\mathbf{q}$ that are orthogonal and have the same magnitude, $|\mathbf{d}| = |\mathbf{c}| = \mathbf{q}$ or $|d| = |c| = q$. Choosing the first row of **M** and its elements d, 0, and 0 respectively to define the related submatrices, minors, and cofactors, the determinant of **M** is computed under the cofactor expansion rule as follows:

$$\begin{aligned} |\mathbf{M}| &= \begin{vmatrix} d & 0 & 0 \\ 0 & c & 0 \\ 0 & 0 & q \end{vmatrix} \\ &= d(+1)|c.q - 0| + 0(-1)|0 - 0| + 0(+1)|0 - 0| \\ &= |d.c.q| \end{aligned}$$

where d, c > or < 0 and q > 0. Since $|d| = |\mathbf{d}|$, $|c| = |\mathbf{c}|$, and $|q| = |\mathbf{q}|$, $|\mathbf{M}|$ is equal to the product $|\mathbf{d}||\mathbf{c}||\mathbf{q}|$ or just $|d.c.q|$. As a 3 × 3 diagonal or scalar matrix, **M** has a nonzero determinant. Also, as **M** is symmetric, so is its determinant: $|\mathbf{M}| = |\mathbf{M}^T|$, which is consistent with the duality and balance features of accounting events.

As with the two-dimensional case before, the determinant |**M**| has similar geometric interpretations. It is equal in absolute terms to the size of the vector space represented by **M**. This vector space is a DE cube, like the one in Fig. 7.2. The location of the DE cube in the three-dimensional vector space depends on the sign (+ or −) of the diagonal elements d and c of **M**. It lies in the first octant (OI) of the three-dimensional space if d, c > 0, that is, if the accounting event is of type 1 and increases the debit and credit accounts alike. It lies in OIII if d, c < 0, that is, if the accounting event of type 1 and decreases the debit and credit accounts alike. If neither is the case, it lies in OII or OIV. Regardless, the determinant |**M**| is computed in the same way. And it is equal to the size of the DE cube represented by **M**.

The determinant of **M** above concerns the DE matrix of an accounting event of type 1. If an accounting event is of type 2, 3, or 4 and has component transactions, it is represented by not a single matrix but a set of discrete matrices (e.g., $\mathbf{M}_1$ and $\mathbf{M}_2$ or $\mathbf{M}_1$ and $\mathbf{M}_4$), one for each component transaction. These discrete matrices are all 3 × 3 diagonal or scalar matrices and have a rank of 3. And their diagonal elements may be positive or negative (+ or −), depending on how the debit and credit accounts change—in the same direction or in different directions. Yet the determinant of the DE matrix of each component transaction is computed as usual. For example, the determinant of $\mathbf{M}_1$ in Sect. 7.1 is computed as

$$|\mathbf{M}_1| = \begin{vmatrix} d_k & 0 & 0 \\ 0 & c_k & 0 \\ 0 & 0 & q_k \end{vmatrix} = |d_k.c_k.q_k|$$

where d_k, c_k > or < 0 and $q_k > 0$. Since $|d_k| = |\mathbf{d}_k|$, $|c_k| = |\mathbf{c}_k|$, and $|q_k| = |\mathbf{q}_k|$, $|\mathbf{M}_1|$ also equals $|\mathbf{d}_k||\mathbf{c}_k||\mathbf{q}_k|$. The determinant of $\mathbf{M}_2$, $\mathbf{M}_3$, and $\mathbf{M}_4$ is computed in the same way. And the magnitude $|\mathbf{M}_1|$ is equal to the size of the DE cube of the component transaction, which is spanned by $\mathbf{d}_k$, $\mathbf{c}_k$, and $\mathbf{q}_k$ as a set of basis vectors.

While the determinant of $\mathbf{M}_1,\ldots,\mathbf{M}_4$ each is computed as usual, the determinants of these matrices cannot be added to get the determinant of $\mathbf{M}$ as a single matrix for the accounting event. Again, this is because of the general property of determinants of matrices: $|\mathbf{A} + \mathbf{B}| \neq |\mathbf{A}| + |\mathbf{B}|$. Given this, the determinant of $\mathbf{M}$ for type 2, 3, or 4 accounting events is computed by, first, constructing a special DE matrix ($\mathbf{M}_S$) with the diagonal elements of $\mathbf{M}_1,\ldots,\mathbf{M}_4$ in absolute value as in the two-dimensional case before. The determinant of $\mathbf{M}_S$ is then computed as usual. Shown below are the special DE matrix of such accounting events and its determinant.

$$\mathbf{M}_S = \begin{bmatrix} \sum|d_k| & 0 & 0 \\ 0 & \sum|c_k| & 0 \\ 0 & 0 & \sum|q_k| \end{bmatrix}$$

$$|\mathbf{M}_S| = \begin{vmatrix} \sum|d_k| & 0 & 0 \\ 0 & \sum|c_k| & 0 \\ 0 & 0 & \sum|q_k| \end{vmatrix} = |\sum|d_k|.\sum|c_k|.\sum q_k$$

where d_k, c_k, and q_k are the debit, credit, and size elements of the kth component transaction and $|\sum|d_k| = \sum|c_k| = \sum q_k$ under the double entry principle. $|\mathbf{M}_S|$ is equal in absolute value to the size of the DE vector space of the accounting event. This space is a cube—a DE cube.

Alternatively, the determinant of the matrix of type 2, 3, or 4 accounting events can be computed by, first, computing the determinant of the DE matrix of each component transaction and the cross products of the main diagonal elements of the DE matrices of the component transactions. And those determinants and cross products are then added up, all in absolute value. The result is equal to the determinant of the DE matrix of the accounting events. If, for example, an accounting event is of type 2 and has two component transactions, thus represented by $\mathbf{M}_1$ and $\mathbf{M}_2$, the determinant of the DE matrix of the event is computed as

$$\mathbf{M}_1 = \begin{bmatrix} d_1 & 0 & 0 \\ 0 & c_1 & 0 \\ 0 & 0 & q_1 \end{bmatrix} \mathbf{M}_2 = \begin{bmatrix} d_2 & 0 & 0 \\ 0 & c_2 & 0 \\ 0 & 0 & q_2 \end{bmatrix}$$

$$\begin{aligned} |\mathbf{M}_S| &= |\mathbf{M}_1| + |\mathbf{M}_2| + |\mathbf{d}_1||\mathbf{c}_1||\mathbf{q}_2| + |\mathbf{d}_2||\mathbf{c}_1||\mathbf{q}_1| + |\mathbf{d}_1||\mathbf{c}_2||\mathbf{q}_1| + |\mathbf{d}_2||\mathbf{c}_2||\mathbf{q}_1| \\ &\quad + |\mathbf{d}_1||\mathbf{c}_2||\mathbf{q}_2| + |\mathbf{d}_2||\mathbf{c}_1||\mathbf{q}_2| \\ &= |d_1.c_1.q_1| + |d_2.c_2.q_2| + |d_1.c_1.q_2| + |d_2.c_1.q_1| + |d_1.c_2.q_1| + |d_2.c_2.q_1| \\ &\quad + |d_1.c_2.q_2| + |d_2.c_1.q_2| \end{aligned}$$

As seen, there are two direct products of the main diagonal elements of $\mathbf{M}_1$ and $\mathbf{M}_2$ and six cross products of the diagonal elements of the two matrices. The determinant $|\mathbf{M}_S|$ for the accounting event is then computed by adding the two determinants $|\mathbf{M}_1|$ and $|\mathbf{M}_2|$ and the six cross products, all in absolute terms. The direct product of the main diagonal elements of $\mathbf{M}_1$ and $\mathbf{M}_2$ respectively equals the size of the DE cube of the related component transaction. And the cross product of the main diagonal elements of $\mathbf{M}_1$ and $\mathbf{M}_2$ equals the size of a rectangular cuboid (or cube) defined by the debit (credit) element of one component transaction and the credit (debit) element of the other. The size of the DE cube of the accounting event equals the size of the two DE cubes combined plus the size of the six rectangular cuboids (or cubes) combined. Computing the size of the DE cube of type 2, 3, or 4 accounting events this way is consistent with the "whole-parts" relation. The DE cube of the accounting event as the "whole" equals the sum of the two component cubes and six cuboids as the "parts." It is also consistent with the notion of linear aggregation in general.

7.3 Accounting Events as 2 × 2 Diagonal Matrices: Examples

The presentation of accounting events in matrix form in the two-dimensional (debit-credit) vector space can be illustrated with the four accounting transaction examples in Sect. 3.2, Chap. 3. The simplest among them is the type 1 accounting event—the sale of goods for \$100 in cash. And the most complex is the type 4 accounting event—the payment of interest on a bond liability, the recognition of interest expense, and the amortization of bond discount, all as one transaction. The type 1 accounting event has no component transaction, and the type 4 accounting event has three component transactions. Shown below are the DE matrices of the four transaction examples.

$$\text{Type 1 event:}\quad \mathbf{M} = \begin{bmatrix} d & 0 \\ 0 & c \end{bmatrix} = \begin{bmatrix} 100 & 0 \\ 0 & 100 \end{bmatrix}$$

$$\text{Type 2 event:}\quad \mathbf{M} = \begin{bmatrix} d_1 + d_2 & 0 \\ 0 & c_1 + c_2 \end{bmatrix} = \begin{bmatrix} 60+40 & 0 \\ 0 & 60+40 \end{bmatrix}$$

$$= \begin{bmatrix} 100 & 0 \\ 0 & 100 \end{bmatrix}$$

$$\text{Type 3 event:}\quad \mathbf{M}_1 = \begin{bmatrix} d_2 & 0 \\ 0 & c \end{bmatrix} = \begin{bmatrix} 40 & 0 \\ 0 & 40 \end{bmatrix} \mathbf{M}_4 = \begin{bmatrix} d_1 & 0 \\ 0 & a \end{bmatrix} = \begin{bmatrix} 60 & 0 \\ 0 & -60 \end{bmatrix}$$

$$\text{Type 4 event:}\quad \mathbf{M}_1 = \begin{bmatrix} d_1 & 0 \\ 0 & c \end{bmatrix} = \begin{bmatrix} 100 & 0 \\ 0 & 100 \end{bmatrix}$$

$$\mathbf{M}_3 = \begin{bmatrix} b & 0 \\ 0 & a_1 \end{bmatrix} = \begin{bmatrix} -120 & 0 \\ 0 & -120 \end{bmatrix}$$

$$\mathbf{M}_4 = \begin{bmatrix} d_2 & 0 \\ 0 & a_2 \end{bmatrix} = \begin{bmatrix} 230 & 0 \\ 0 & -230 \end{bmatrix}$$

All matrices above are a 2 × 2 diagonal matrix, and the subscripts 1, 3, and 4 denote the quadrant of the two-dimensional debit-credit vector space where each matrix is presented geometrically. **M**, $\mathbf{M}_1$, and $\mathbf{M}_3$ in particular are a 2 × 2 scalar matrix as their main diagonal elements are the same scalar number, positive or negative. The columns of each matrix represent the debit and credit dimensions respectively of the two-dimensional vector space. The first (second) row of each matrix represents the debit (credit) element d (c) of each accounting event or each component transaction. The first matrix concerns the type 1 accounting event, and the first (second) diagonal element 100 represents the cash (sales revenue) \$100 as the debit (credit) entry. The second matrix concerns the type 2 accounting event—the sale of goods for \$100, receiving \$60 in cash and holding \$40 on account. As the event has two component transactions, each diagonal element of **M** is made up of two components, 60 and 40. The next two matrices $\mathbf{M}_1$ and $\mathbf{M}_4$ as a pair represent the type 3 accounting event—the purchase of goods, partly in cash and partly on account. $\mathbf{M}_1$ represents the account-payable part of the event, and $\mathbf{M}_4$ the cash part. The credit entry is shown by two numbers: 40 for the account payable as an increase and − 60 for the cash as a decrease. Likewise, the debit entry is also shown by two numbers: $d_1 = 60$ and $d_2 = 40$, which together represent the inventory asset as an increase. The next three matrices $\mathbf{M}_1$, $\mathbf{M}_3$, and $\mathbf{M}_4$ as a set represent the type 4 accounting event. $\mathbf{M}_1$ represents the amortization of the bond discount \$100 as an increase in the bond liability account, $\mathbf{M}_3$ the payment of the unpaid interest \$120 as a decrease in the interest payable account, and $\mathbf{M}_4$ the payment of the current-period interest expense \$230 as an increase. Some main diagonal elements of those matrices are negative numbers as they mean a decrease in the related account. Yet the total debits of the accounting event in absolute terms are equal to the total credits under the double entry principle: $100 + |-120| + 230 = 100 + |-120| + |-230|$.

Shown below is the determinant of each DE matrix above. The determinant of **M** for the type 1 and 2 accounting events is computed as usual. For the type 3 and 4 accounting events, a special DE matrix ($\mathbf{M}_S$) is constructed as discussed before, and its determinant is then computed as usual.

$$\begin{vmatrix} 100 & 0 \\ 0 & 100 \end{vmatrix} = \left|100^2 - 0\right| = 100^2$$

$$\begin{vmatrix} 60 + 40 & 0 \\ 0 & 60 + 40 \end{vmatrix} = \left|100^2 - 0\right| = 100^2$$

$$\begin{vmatrix} 40 + 60 & 0 \\ 0 & 40 + |-60| \end{vmatrix} = \left|100^2 - 0\right| = 100^2$$

$$
\begin{vmatrix} 100 + |-120| + 230 & 0 \\ 0 & 100 + |-120| + |-230| \end{vmatrix}
$$
$$
= |(100 + |-120| + 230)(100 + |-120| + |-230|) - 0.0|
$$
$$
= 450^2
$$

For the type 3 and 4 accounting events, $|\mathbf{M}_S|$ is not equal to the sum of the determinants of the discrete DE matrices of their component transactions. For the type 3 event, for example, $|\mathbf{M}_S| \neq |\mathbf{M}_1| + |\mathbf{M}_4|$. Again, this is because of the general property of determinants of matrices: $|\mathbf{A} + \mathbf{B}| \neq |\mathbf{A}| + |\mathbf{B}|$.

Geometrically, the determinant $|\mathbf{M}|$ or $|\mathbf{M}_S|$ in absolute value is equal to the size of the DE vector space represented by the matrix. The location of this vector space in the two-dimensional (debit-credit) vector space depends on the sign (+ or −) of the diagonal elements of the matrix. The DE vector space of the type 1 and 2 accounting events lies in QI of the vector space since these two events increase the debit and credit accounts alike. For the type 3 accounting event, the determinant $|\mathbf{M}_S|$ is equal to the combined size of the DE vector spaces represented by $\mathbf{M}_1$ and $\mathbf{M}_4$ plus the vector spaces defined by the debit and credit elements of the component transactions. And the DE vector space represented by $|\mathbf{M}_S|$ lies across QI and QIV. The determinant $|\mathbf{M}_S|$ for the type 4 accounting event is computed similarly, by adding $|\mathbf{M}_1|$, $|\mathbf{M}_3|$, $|\mathbf{M}_4|$ and the cross products of the debit and credit elements of the component transactions. The DE vector spaces of the three component transactions lie in different quadrants—one in QI, QIII, and QIV respectively. And the DE vector space of the accounting event lies across the four quadrants QI to QIV.

7.4 Accounting Events as 3 × 3 Diagonal Matrices: Examples

The DE matrices, and their determinants, of the four accounting transaction examples in the previous section can also be examined in the three-dimensional (debit-credit-size) vector space. All those matrices are now a 3 × 3 diagonal or scalar matrix and have a rank of 3. Shown below are the DE matrices of the four accounting transaction examples before.

$$
\text{Type 1 event: } \mathbf{M} = \begin{bmatrix} d & 0 & 0 \\ 0 & c & 0 \\ 0 & 0 & q \end{bmatrix} = \begin{bmatrix} 100 & 0 & 0 \\ 0 & 100 & 0 \\ 0 & 0 & 100 \end{bmatrix}
$$

$$\text{Type 2 event:} \quad \mathbf{M} = \begin{bmatrix} d_1 + d_2 & 0 & 0 \\ 0 & c_1 + c_2 & 0 \\ 0 & 0 & q_1 + q_2 \end{bmatrix}$$

$$= \begin{bmatrix} 60 + 40 & 0 & 0 \\ 0 & 60 + 40 & 0 \\ 0 & 0 & 60 + 40 \end{bmatrix}$$

$$\text{Type 3 event:} \; \mathbf{M}_1 = \begin{bmatrix} d_2 & 0 & 0 \\ 0 & c & 0 \\ 0 & 0 & q_2 \end{bmatrix} = \begin{bmatrix} 40 & 0 & 0 \\ 0 & 40 & 0 \\ 0 & 0 & 40 \end{bmatrix}$$

$$\mathbf{M}_4 = \begin{bmatrix} d_1 & 0 & 0 \\ 0 & a & 0 \\ 0 & 0 & q_1 \end{bmatrix} = \begin{bmatrix} 60 & 0 & 0 \\ 0 & -60 & 0 \\ 0 & 0 & 60 \end{bmatrix}$$

$$\text{Type 4 event:} \; \mathbf{M}_1 = \begin{bmatrix} d_1 & 0 & 0 \\ 0 & c & 0 \\ 0 & 0 & q_1 \end{bmatrix} = \begin{bmatrix} 100 & 0 & 0 \\ 0 & 100 & 0 \\ 0 & 0 & 100 \end{bmatrix}$$

$$\mathbf{M}_3 = \begin{bmatrix} b & 0 & 0 \\ 0 & a_1 & 0 \\ 0 & 0 & q_2 \end{bmatrix} = \begin{bmatrix} -120 & 0 & 0 \\ 0 & -120 & 0 \\ 0 & 0 & 120 \end{bmatrix}$$

$$\mathbf{M}_4 = \begin{bmatrix} d_2 & 0 & 0 \\ 0 & a_2 & 0 \\ 0 & 0 & q_3 \end{bmatrix} = \begin{bmatrix} 230 & 0 & 0 \\ 0 & -230 & 0 \\ 0 & 0 & 230 \end{bmatrix}$$

The subscripts 1, 3, and 4 of the matrices denote the octant of the three-dimensional vector space where the DE vector space of each component transaction lies. All those DE matrices are a diagonal matrix, and $\mathbf{M}$ and $\mathbf{M}_1$ in particular are a scalar matrix. The first DE matrix represents the type 1 accounting event, which increases the cash account and the sales revenue account alike. The second DE matrix represents the type 2 accounting event, which has two component transactions and increases the debit and credit accounts alike. The next two matrices $\mathbf{M}_1$ and $\mathbf{M}_4$ represent the type 3 accounting event. This event has two credit accounts, the cash \$60 as a decrease and the account payable \$40 as an increase. And the diagonal element -60 of $\mathbf{M}_4$ represents the cash paid. The next three matrices $\mathbf{M}_1$, $\mathbf{M}_3$, and $\mathbf{M}_4$ represent the type 4 accounting event, which has three component transactions. $\mathbf{M}_1$ is a scalar matrix, and the other two are diagonal matrices.

Each DE matrix above represents the DE vector space (cube) of the related accounting event or the related component transaction. The location of the respective DE cube in the three-dimensional vector space depends on the sign (+ or −) of the diagonal elements of the DE matrix. The DE cube of the first two accounting events lies in the first octant (OI) of the vector space. The third accounting event has the two credit accounts, but these two accounts do not change in the same direction. Thus, one of the two component DE cubes lies in OI, and the other in OIV. And the DE cube of the accounting event lies across OI and OIV. The DE cube of the fourth accounting event lies across the four octants of the three-dimensional vector space since this event has three component transactions and the multiple accounts on the debit and credit sides respectively do not change in the same direction. The DE cubes of the component transactions lie in three different octants, one in OI, OIII, and OIV respectively.

Shown below is the determinant of the DE matrix of each accounting event above (in the same order before).

$$\begin{vmatrix} 100 & 0 & 0 \\ 0 & 100 & 0 \\ 0 & 0 & 100 \end{vmatrix} = 100^3$$

$$\begin{vmatrix} 60+40 & 0 & 0 \\ 0 & 60+40 & 0 \\ 0 & 0 & 60+40 \end{vmatrix} = 100^3$$

$$\begin{vmatrix} 40+60 & 0 & 0 \\ 0 & 40+|-60| & 0 \\ 0 & 0 & 40+60 \end{vmatrix} = 100^3$$

$$\begin{vmatrix} 100+|-120|+230 & 0 & 0 \\ 0 & 100+|-120|+|-230| & 0 \\ 0 & 0 & 100+120+230 \end{vmatrix} = 450^3$$

The determinant of the DE matrix of the type 3 and 4 accounting events is computed by constructing a special DE matrix ($\mathbf{M}_S$) for the reason explained in Sect. 7.2. For the type 2, 3, and 4 accounting events, the determinant $|\mathbf{M}|$ or $|\mathbf{M}_S|$ is not the same as the sum of the determinants of the DE matrices of their component transactions. For the type 3 event, for example, $|\mathbf{M}_S| \neq |\mathbf{M}_1| + |\mathbf{M}_4|$. Again, this is because of the general property of the determinants of matrices: $|\mathbf{A} + \mathbf{B}| \neq |\mathbf{A}| + |\mathbf{B}|$.

When seen in the three-dimensional vector space, the determinant of each DE matrix above is equal to the size of the DE vector space (cube) represented by the matrix. The location of the DE cube in the three-dimensional vector space depends on the sign (+ or −) of the diagonal elements of the matrix. As for the type 1 accounting event, for example, the three diagonal elements are all positive so that the DE cube represented by **M** lies in OI. The DE cube represented by $\mathbf{M}_S$ for the type 4 accounting event lies across the four octants OI to OIV.

7.5 Features of the Double Entry Matrix

The DE matrices of accounting events discussed in the previous sections have interesting features. Some of these features were discussed earlier (in Sect. 7.1). For example, **M** is a symmetric matrix so that it has the features of symmetric matrices in general. It is the same as its transpose: $\mathbf{M} = \mathbf{M}^T$. As such, interchanging the rows and columns of **M** does not change its determinant: $|\mathbf{M}| = |\mathbf{M}^T|$. Nor does it change the vector space represented by the matrix. As a diagonal or scalar matrix, **M** has a nonzero determinant. Hence, it is nonsingular and has an inverse. Also, by nature, **M** cannot be a null matrix since no accounting event can take place with a zero monetary size (volume). Nor can it be a rectangular matrix because the double entry principle does not allow it to be. The columns (vectors) of **M** are mutually orthogonal. The diagonal elements of **M** have the same magnitude under the double entry principle although their signs can differ (+ or −).

Besides those features above, the DE matrix of accounting events also has some other features. These are conceptual features—duality and balance, additivity, symmetry, invertibility, and orthogonality respectively.

Duality and Balance As a diagonal or scalar matrix, the DE matrix **M** of accounting events reflects the duality and balance features of the accounting events or, more generally, the features and principles of double entry accounting. Accounting events (transactions), like the economic (capital) resources of a business entity, have duality and balance as their inherent features (see Sect. 2.1, Chap. 2). As such, their dual elements, debit and credit, always co-exist, co-vary, and do so with balance. **M** reflects these duality and balance features of accounting events and the contrasting yet symmetric nature of double entries. This is true, regardless of whether **M** is a diagonal or scalar matrix. If **M** is a scalar matrix in particular, it can be written as $\mathbf{M} = k\mathbf{I}$, where k is a scalar number and **I** is a unit matrix. And the unit matrix **I** is considered to reflect the duality and balance features of accounting events. Interchanging the rows and columns of **M** does not change its features and properties. Nor does it change the determinant of **M** or the size of the DE vector space (square or cube) represented by **M**; only the interpretations differ.

Given the above, **M** is not just any diagonal matrix but a special kind of diagonal matrix. The DE vector space represented by **M** is not just any vector space but a special kind of vector space, square or cube, as discussed in Sect. 3.3, Chap. 3, and Sect. 4.3, Chap. 4. The determinant of **M** in absolute terms is equal to the size of the DE vector space. It tells how large this DE vector space is relative to the vector space represented by a unit matrix, which is equal to 1.

Additivity Mathematically, matrices can be added as long as they are conformable for addition, regardless of the signs (+ and −) of their elements. And the resultant matrix still has a meaning. This, however, is not always true for the DE matrices of accounting events. First of all, the notion of additivity does not apply to **M** for type 1 accounting events, since such accounting events have no component transactions. Thus, there are no discrete DE matrices to be added to get **M** as a single DE matrix.

The additivity of matrices applies to **M** for type 2, 3, or 4 accounting events, which have component transactions and are represented by a set of discrete DE matrices. Each of these component transactions has the DE matrix of its own. For such accounting events, the addition of those discrete DE matrices works as usual if they change the multiple accounts on the debit and credit sides respectively in the same direction. If they do not, the usual matrix additivity rule does not work since adding the discrete DE matrices of the component transactions does not give the DE matrix of the accounting events. That is, those discrete DE matrices cannot be linearly added element by element in order to get **M** as a single DE matrix, although it can be done mathematically. If they are added, the resultant matrix is not a DE matrix since $\sum d_k \neq \sum c_k$ or $\sum \mathbf{M}_k \neq \mathbf{M}$, which is not allowed under the double entry principle yet is fine mathematically. $\sum \mathbf{M}_k$ as the sum of the "parts" is not the same as **M** as the "whole." The sum does not represent an accounting event. Nor does it reflect the duality and balance features of the accounting event and the contrasting yet symmetric nature of double entries. Also, the sum of the determinants of the discrete DE matrices is not equal to the determinant of **M** ($\sum|\mathbf{M}_k| \neq |\mathbf{M}|$). This was the reason why the special matrix $\mathbf{M}_S$ was constructed for the type 3 and 4 accounting transaction examples before, when the determinant of their DE matrices was computed.

Symmetry Reflecting the duality and balance features of accounting events, **M** is always a symmetric matrix in a broad sense. Thus, it is the same as its transpose ($\mathbf{M} = \mathbf{M}^T$). Interchanging its rows and columns does not change its features. Nor does this affect its determinant ($|\mathbf{M}| = |\mathbf{M}^T|$). And the size of the DE vector space represented by **M** also remains the same. What this means is that the inverse of the determinant of **M** is equal to the determinant of the inverse of **M** ($|\mathbf{M}|^{-1} = |\mathbf{M}^{-1}|$). And the product of **M** and its inverse is a unit matrix: $\mathbf{M}\mathbf{M}^{-1} = \mathbf{M}^{-1}\mathbf{M} = \mathbf{I}$. This unit matrix reflects the duality and balance features of accounting events and the contrasting yet symmetric nature of the double entry principle. And the determinant of $\mathbf{M}\mathbf{M}^{-1}$ is 1, and the size of the vector space represented by $\mathbf{M}\mathbf{M}^{-1}$ is also 1. That is, $|\mathbf{M}\mathbf{M}^{-1}| = |\mathbf{M}^{-1}\mathbf{M}| = |\mathbf{I}| = 1$.

Invertibility Diagonal matrices in general have a nonzero determinant. Thus, they are nonsingular and invertible and, hence, have an inverse. The main diagonal elements of the inverse matrix are the reciprocal of the corresponding main diagonal elements of the original matrix. This is also true for **M**. As a 2 × 2 or 3 × 3 diagonal or scalar matrix, **M** has a nonzero determinant ($|\mathbf{M}| \neq 0$). Thus, it has an inverse ($\mathbf{M}^{-1}$), which is also a 2 × 2 or 3 × 3 diagonal or scalar matrix. The main diagonal elements of $\mathbf{M}^{-1}$ are the reciprocal of the corresponding diagonal elements of **M**. Also, the determinant of the inverse matrix $\mathbf{M}^{-1}$ is equal to the inverse of the determinant of **M**: $|\mathbf{M}^{-1}| = |\mathbf{M}|^{-1}$.

In matrix algebra, the inverse of matrices is found by performing a series of algebraic operations under the cofactor expansion rule, also called the Laplace expansion formula. The inverse ($\mathbf{M}^{-1}$) of $\mathbf{M}$ can also be found by following the same procedure under the same rule as shown below.

$$\mathbf{M}^{-1} = \mathbf{C}^{T}/|\mathbf{M}| = \begin{bmatrix} c & 0 \\ 0 & d \end{bmatrix} / |d.c| = \begin{bmatrix} 1/d & 0 \\ 0 & 1/c \end{bmatrix}$$

or

$$\mathbf{M}^{-1} = \mathbf{C}^{T}/|\mathbf{M}| = \begin{bmatrix} c.q & 0 & 0 \\ 0 & d.q & 0 \\ 0 & 0 & d.c \end{bmatrix} / |d.c.q|$$

$$= \begin{bmatrix} 1/d & 0 & 0 \\ 0 & 1/c & 0 \\ 0 & 0 & 1/q \end{bmatrix}$$

where $\mathbf{C}^{T}$ is the transpose of the cofactor matrix $\mathbf{C}$ of $\mathbf{M}$, and $|\mathbf{M}|$ is the determinant of $\mathbf{M}$.[4]

The inverse matrix $\mathbf{M}^{-1}$ above concerns $\mathbf{M}$ for accounting events of type 1, having no component transactions. If an accounting event is of type 2, 3, or 4, the process of finding $\mathbf{M}^{-1}$ is not as simple since they have component transactions so that there are multiple DE matrices to deal with. If the multiple accounts of the accounting event on the debit or credit side (both sides) change in the same direction (increase or decrease), $\mathbf{M}^{-1}$ is found as shown above since $\sum \mathbf{M}_k = \mathbf{M}$ in this case. If the multiple accounts on each side do not change in the same direction, the special DE matrix $\mathbf{M}_S$ is constructed, first, and its inverse ($\mathbf{M}_S^{-1}$) is then found as usual according to the cofactor expansion rule.

Once $\mathbf{M}^{-1}$ or $\mathbf{M}_S^{-1}$ is found, some additional features of the DE matrix can be explored. For example, the product of $\mathbf{M}$ (or $\mathbf{M}_S$) and its inverse yields a unit matrix. Also, the transpose of $\mathbf{M}^{-1}$ is equal to the inverse of $\mathbf{M}^{T}$. If, for example, $\mathbf{M}$ is a 2×2 diagonal or scalar matrix,

$$\mathbf{M}\mathbf{M}^{-1} = \mathbf{I} \rightarrow \begin{bmatrix} d & 0 \\ 0 & c \end{bmatrix} \times \begin{bmatrix} 1/d & 0 \\ 0 & 1/c \end{bmatrix} = \begin{bmatrix} 1 & 0 \\ 0 & 1 \end{bmatrix}$$

[4]The cofactor matrix here means a matrix of cofactors. How the cofactor expansion rule works for the computation of the determinant of matrices was explained in Sect. 7.2. The process of finding the inverse of matrices under this rule is slightly different from the process of computing the determinant of matrices using the same rule as shown above.

$$\left(\mathbf{M}^{-1}\right)^{\mathrm{T}} = \left(\mathbf{M}^{\mathrm{T}}\right)^{-1} = \begin{bmatrix} 1/\mathrm{d} & 0 \\ 0 & 1/\mathrm{c} \end{bmatrix} = \mathbf{M}^{-1}$$

Similarly, the following also hold for the special DE matrix $\mathbf{M}_{\mathrm{S}}$:

$$\mathbf{M}_{\mathrm{S}}\mathbf{M}_{\mathrm{S}}^{-1} = \mathbf{I} \rightarrow \begin{bmatrix} \sum|\mathrm{d}_k| & 0 \\ 0 & \sum|\mathrm{c}_k| \end{bmatrix} \times \begin{bmatrix} 1/\sum|\mathrm{d}_k| & 0 \\ 0 & 1/\sum|\mathrm{c}_k| \end{bmatrix} = \begin{bmatrix} 1 & 0 \\ 0 & 1 \end{bmatrix}$$

$$\left(\mathbf{M}_{\mathrm{S}}^{-1}\right)^{\mathrm{T}} = \left(\mathbf{M}_{\mathrm{S}}^{\mathrm{T}}\right)^{-1} = \begin{bmatrix} 1/\sum|\mathrm{d}_k| & 0 \\ 0 & 1/\sum|\mathrm{c}_k| \end{bmatrix} = \mathbf{M}_{\mathrm{S}}^{-1}$$

Orthogonality A square matrix is orthogonal if its columns (or rows) are orthogonal vectors with a magnitude equal to a unity (1). Alternatively, a matrix is orthogonal if its determinant is 1, its transpose is equal to its inverse, or the product of the matrix and its transpose yields a unit matrix. That is, if a matrix (**A**) is an orthogonal matrix, $|\mathbf{A}| = 1$, $\mathbf{A}^{\mathrm{T}} = \mathbf{A}^{-1}$, and $\mathbf{A}\mathbf{A}^{\mathrm{T}} = \mathbf{A}^{\mathrm{T}}\mathbf{A} = \mathbf{I}$. And the determinant of $\mathbf{A}\mathbf{A}^{\mathrm{T}}$ is 1 as the determinant $|\mathbf{I}|$ is 1. As such, an orthogonal matrix is not just any square matrix but a special kind of square matrix whose rows or columns are orthogonal vectors with a unit length and whose determinant equals the size of the vector space represented by **I**, which is 1.

The DE matrix **M** is a square matrix whose column vectors **d** and **c**, or **d**, **c**, and **q**, are orthogonal as they are coordinate vectors in the two-dimensional (debit-credit) or three-dimensional (debit-credit-size) vector space. Also, their dot product (e.g., $\mathbf{d}.\mathbf{c}^{\mathrm{T}}$) is zero (0), meaning that those vectors are mutually orthogonal ($\theta = 90°$). However, **M** is not an orthogonal matrix since it does not have the features of an orthogonal matrix indicated above. The row (or column) vectors of **M** do not typically have a unit length unless the accounting event takes place with d = c = \$1, for example, which is likely rare in reality. As such, most likely, the determinant of **M** is not equal to 1. Nor is the transpose of **M** likely to be the same as its inverse ($\mathbf{M}^{\mathrm{T}} \neq \mathbf{M}^{-1}$). Further, the product of **M** and $\mathbf{M}^{\mathrm{T}}$ does not likely yield a unit matrix. These are shown below.

$$|\mathbf{M}| = \begin{vmatrix} \mathrm{d} & 0 \\ 0 & \mathrm{c} \end{vmatrix} = |\mathrm{d.c} - 0.0| \neq 1$$

$$\mathbf{M}^{\mathrm{T}} \neq \mathbf{M}^{-1}$$

$$\begin{bmatrix} \mathrm{d} & 0 \\ 0 & \mathrm{c} \end{bmatrix} \neq \begin{bmatrix} 1/\mathrm{d} & 0 \\ 0 & 1/\mathrm{c} \end{bmatrix}$$

$$\mathbf{M}\mathbf{M}^{\mathrm{T}} \neq \mathbf{I}$$

$$\begin{bmatrix} \mathrm{d} & 0 \\ 0 & \mathrm{c} \end{bmatrix} \times \begin{bmatrix} \mathrm{d} & 0 \\ 0 & \mathrm{c} \end{bmatrix} = \begin{bmatrix} \mathrm{d}^2 & 0 \\ 0 & \mathrm{c}^2 \end{bmatrix} \neq \begin{bmatrix} 1 & 0 \\ 0 & 1 \end{bmatrix}$$

The Aggregation of Monetary Measures in Accounting

8

From a practical standpoint, the process of accounting consists of four sequential activities: measuring, recording, aggregating (summarizing), and reporting. The most important among these activities is the second step *recording*—the recording of accounting transactions (events) according to the double entry principle. This step is at the heart of accounting and makes double entry accounting so unique. It was for this reason why all previous chapters focused on the recording step of the accounting process, namely the presentation of accounting events in geometric and vector spaces according to the double entry principle. This chapter examines the next step *aggregating* of the accounting process. It discusses how to aggregate monetary measures in accounting in geometric and vector spaces yet according to the double entry principle. The geometric setting used to do so is a two-dimensional debit-credit Cartesian space. A three- or higher-dimensional geometric setting is not relevant since the aggregations of monetary measures at different levels in accounting take place with respect to the debits and credits of individual accounts and groups of accounts.

8.1 Measures and Measurements in Accounting

When viewed conceptually, a measure represents a measured quantity of an object, a thing, or a property. It is shown as a number. Measurement is an act of taking a measure and showing it as a number. Once generated, the measures are then aggregated into another measure, which is also shown as a number. Here, aggregation means a linear addition of multiple measures scaled by a factor of 1, not just any scale factor. It means reducing the multiple measures or numbers to a single measure or number. Algebraically, it is like $A = \sum a_k$, where A is an aggregate and a_k is a measure. In this regard, the notion of aggregation is compatible with the "whole-parts" relation in the previous chapters. The aggregate A is considered the "whole" and the measure a_k is considered a "part." In order for the "whole-parts" relation to hold, there should be no interaction among the measures or "parts" being aggregated.

© The Author(s), under exclusive license to Springer Nature Switzerland AG 2025

B. T. Ro, *The Geometry of Accounting*, Springer Texts in Business and Economics,
https://doi.org/10.1007/978-3-031-88892-2_8

The measures being aggregated should be independent. The independence of measures is necessary to ensure the "whole-parts" relation to hold. The result of aggregation is another measure, which is called a summary measure, an aggregate, a sum, a total, or even the "whole."

In order to get a meaningful aggregate, not only should the individual measures being aggregated be independent, but they should also be homogeneous. In other words, the measures being aggregated should represent a common attribute of objects or things, tangible or intangible, which can be measured and shown in the same measurement unit. Otherwise, the resultant aggregate has no meaningful interpretation.

Before an aggregation process begins, there must exist a set of measures, first. They are the initial measures of the quantity or property of objects or things, whether tangible or intangible. They are primitive measures and shown as numbers. The word *primitive* literally means "initial," "original," or "primary." The aggregation of primitive measures yields a new measure—an aggregate, which is a derived measure. This measure is also a number. The word *derived* means "resultant" or "secondary." As such, all derived measures originate from primitive measures of some kind. These derived measures can be aggregated further in order to get a higher level of derived measures.

From a practical standpoint, the process of accounting consists of four sequential activities—measuring, recording, aggregating, and reporting. In order for the process to begin, something to measure must exist, first. This is the monetary measure of individual accounts. These accounts are the debits and credits respectively of accounting events and come to exist when the events take place and are recorded under the double entry principle during each accounting period. A monetary measure, whether debit or credit, of each account is a primitive measure and shown by a number. Determining the monetary measures of the debits and credits of individual accounts and showing them as numbers is an act of measurement in accounting. These numbers are then aggregated into a derived measure, which is called the total debits or total credits of those accounts.

To get total debits and total credits respectively of individual accounts as meaningful aggregates, accounting events and individual accounts must have a common (homogeneous) attribute that can be measured and shown in the same measurement unit. In accounting, this attribute is the monetary value of the things that are interchanged in accounting events and recorded as the debits and credits of the events. The measurement unit used to measure their monetary values is a monetary unit—the same monetary unit (e.g., US dollar or euro, not a mix of both). The monetary measure of the debit and credit respectively of each account is a primitive measure. An aggregation in accounting begins with the debits and credits of individual accounts as primitive measures. And all aggregations begin and end at the end of each accounting period.

8.2 Aggregating Monetary Measures in Accounting

The aggregation of monetary measures, also called accounting measures, in accounting is performed at multiple levels in a sequential order, all at the end of each accounting period. At the first level, it is performed with respect to the primitive monetary measures, debits and credits, of individual accounts, not accounting events. All subsequent aggregates originate from those primitive measures. Once aggregated, the primitive measures become derived measures. These derived measures are generated at different levels as the aggregation of monetary measures proceeds at different levels. What is reported in the final summary reports—the financial statements—is the ultimate derived measures (aggregates) of individual accounts or groups of similar accounts such as total assets, total liabilities, and total equity respectively. All monetary measures, whether primitive or derived, are nonnegative in substance as meant by the nonnegativity-of-money convention in accounting (see Sect. 2.2, Chap. 2). This is true even for monetary measures as decreases in individual accounts.

In order to get a meaningful aggregate at each level of aggregation, the monetary measures of the debits and credits of individual accounts being aggregated should be independent. That is, they should come from different accounting events or relate to different accounts. There should be no interaction among the monetary measures being aggregated at each level since aggregation is a linearly additive concept as indicated before. The independence of monetary measures is necessary to ensure $A = \sum a_k$), for example. The existence of any interaction among the monetary measures being aggregated can make A differ from $\sum a_k$.

Level 1 Aggregation The first level of the aggregation of monetary measures in accounting is performed to get total debits and total credits respectively account by account at the end of each accounting period. This is done by adding the debits and credits respectively of each account as primitive measures. These primitive measures come to exist as accounting events involving the account take place and are recorded during the accounting period. For each account, the debits are homogeneous since they are either increases or decreases in monetary quantity, not a mix of both. So are the credits. Once aggregated, the total debits and total credits respectively of the account become total increases or total decreases, depending on the nature of the account (e.g., asset or liability). They are a pair of level 1 aggregates, or level 1 derived measures, of the account. The debits being aggregated are compatible and homogeneous in that they all are the debits of the same account as increases or decreases, not a mix of both. Also, they are independent as they come from different accounting events. There is no interaction among the debits. As such, they can be linearly added to get total debits as an aggregate. The same is also true for the credits being aggregated; they are all increases or decreases, not a mix of both, and compatible, homogeneous, and independent so that they can be linearly added to get total credits as an aggregate. As for the cash account, for example, the debits (credits) are all increments (decrements) so that they are compatible and homogeneous. And they are independent as they come from different cash transactions.

Thus, they can be linearly added as the total debits (total credits) of the cash account, which are total increases (total decreases). This aggregate is a level 1 aggregate of the cash account. It is equivalent to the "whole" in the "whole-parts" relation indicated before, while each debit (credit) entry of the cash account is a "part."

For each account, regardless of its kind or nature (e.g., asset or liability), the level 1 aggregation of the primitive debits and primitive credits respectively is shown algebraically as

$$D_T = \sum d_k$$
$$C_T = \sum c_k$$

where D_T and C_T denote the total debits and total credits respectively of the account, and d_k and c_k are its kth debit amount and k^{th} credit amount, respectively. These debit and credit amounts are primitive measures, and D_T and C_T are derived measures. Whether D_T or C_T means total increases or total decreases depends on the nature of the account—asset or liability, or expense or revenue, for example. Regardless, D_T and C_T are the level 1 aggregates, or level 1-derived measures, of the account. They are nonnegative (D_T, $C_T \geq 0$) since d_k or c_k is nonnegative (d_k, $c_k \geq 0$). No accounting event can take place with d_k and c_k as a negative or zero monetary quantity. Also, by nature, a monetary quantity in general cannot be intrinsically negative, as meant by the nonnegativity-of-money convention in accounting.

Some accounts do not have primitive debits or primitive credits during an accounting period. Thus, $D_T = 0$ or $C_T = 0$. For revenue accounts, for example, the primitive measure d_k is zero ($d_k = 0$) and $D_T = 0$ unless there is an adjustment or a correction on their debit side. Similarly, for expense accounts, c_k is zero ($c_k = 0$) and $C_T = 0$ unless there is an adjustment or a correction on their credit side. For some other accounts, $D_T = 0$ or $C_T = 0$ for other reasons. If, for example, a firm has an account receivable but has no collection of it during the accounting period, the account has no credit entry. Thus, $c_k = 0$ and $C_T = 0$.

For each account, the level 1 aggregation of its debits and credits respectively yields one of the three possible results:

$$D_T >, =, \text{or} < C_T$$

For most accounts, the case $D_T >$ or $< C_T$ is more typical in reality than is the case $D_T = C_T$.

Level 2 Aggregation This level of aggregation is also carried out account by account in order to get the period-end balance of each account. It is performed in two steps. In the first step, D_T is offset against C_T, or vice versa, in order to get the net change (D_C or C_C) in the account during the accounting period:

$$D_C = D_T - C_T$$

or

$$C_C = C_T - D_T$$

where D_C and C_C mean the net change—increase or decrease—in the account, depending on the nature of the account (e.g., asset or liability). Every account has either D_C or C_C, not both.

For revenue accounts, $D_T = 0$ unless there is an adjustment or a correction (d_k) on their debit side. They cannot have a debit entry as a primitive measure. Thus, $C_C = C_T$. For expense accounts, $C_T = 0$ unless there is an adjustment or a correction (c_k) on their credit side. They cannot have a credit entry as a primitive measure. Thus, $D_C = D_T$. For balance sheet accounts, D_C is obtained by subtracting C_T from D_T or C_C is obtained by subtracting D_T from C_T, depending on the nature of the accounts (e.g., asset or liability).

For each account, whether it is a balance sheet account or an income statement account, the net change is

$$D_C >, =, \text{or} < 0$$

or

$$C_C >, =, \text{or} < 0$$

For most accounts, it is typical that the net change D_C or C_C is positive. For some accounts, however, it can be zero. For example, a firm can have a short-term asset such as an account receivable or a note receivable recorded in the accounting period and collect it by the end of the period. The same is also true for a short-term liability such as an account payable or a note payable. Also, D_C or C_C can be negative for some accounts. If, for example, a firm has an account receivable from the previous accounting period and collects it during this period, with no new account receivable, the net change D_C will be negative. For most accounts, however, it is likely rare that D_C or $C_C < 0$ unless there is an adjustment or a correction in the accounts. This is true for income statement (revenue and expense) accounts in particular.

In the second step of level 2 aggregation, D_C or C_C for each account is added to the existing (period-beginning) balance (B_B) of the same account as shown below. The result is the period-end balance (E_B) of the account, which is a level 2 aggregate.

$$\begin{aligned} E_B &= B_B \pm D_C \quad \text{(asset and expense accounts)} \\ &= B_B \pm C_C \quad \text{(liability, equity, and revenue accounts)} \end{aligned}$$

For most accounts, $E_B \geq 0$ as it is rare that $B_B + D_T < C_T$ for asset and expense accounts and $B_B + C_T < D_T$ for liability, equity, and revenue accounts (unless there is a large adjustment or correction of the account). For income statement accounts, $B_B = 0$ since they are closed at the end of each accounting period, in order to compute the net income of the period. Thus, $E_B = C_T = C_C$ for revenue accounts, and $E_B = D_T = D_C$ for expense accounts. And their period-end balance is positive ($E_B > 0$) since their primitive (debit or credit) measures cannot be zero or negative. For balance sheet accounts, $E_B \geq 0$. For some of these accounts, their period-end balance may be zero ($E_B = 0$). If, for example, a firm fully collects an account receivable by the end of the accounting period, $E_B = 0$ even if $B_B > 0$. The same is also true if a firm fully pays an account payable by the end of the accounting period.

The period-end balance E_B of each account is reported as the balance of the account on the general trial balance at the end of each accounting period. For some accounts, it is also reported in the year-end financial statements (e.g., cash, inventory, and property, plant, and equipment).

Level 3 Aggregation This level of aggregation is performed by adding the level 2 aggregates of different accounts in the same group or in a similar class in order to get various group totals. For balance sheet accounts, it is performed to get group totals such as current or total assets, current or total liabilities, and total owner equity. For income statement accounts, it is performed to get total revenues and total expenses, respectively. For these accounts, the level 3 aggregation continues on in order to compute the net income of the accounting period and include it in the retained earnings account, which is part of total owner equity. To determine the net income, total expenses and total revenues respectively of the accounting period are first determined as separate level 3 aggregates. The former is then subtracted from the latter in order to get the net income of the accounting period. This is done in similar ways the net change C_C for the retained earnings account is computed in the first step of level 2 aggregation. As a net change, the net income is then added to C_C for the retained earnings account and included in the period-end balance E_B of the account.

What is reported in the financial statements and shown in the fundamental accounting model at the end of each accounting period is the ultimate level 3 aggregates of all accounts, grouped as total assets, total liabilities, and total owner equity. Each of these aggregates is computed as

$$\text{Assets} = \sum E_{Bi} \quad (i = \text{asset } i)$$

$$\text{Liabs.} = \sum E_{Bi} \quad (i = \text{liability } i)$$

$$\text{Equity} = \sum E_{Bi} \quad (i = \text{equity account } i)$$

The three level 3 aggregates are ultimate aggregates since they cannot be aggregated any further. Also, they are positive. They cannot even be zero since no business entity (firm) can have zero total assets or zero total owner equity or, more generally, zero capital (economic) resources.

8.3 The Geometry of Aggregations in Accounting

The three levels of the aggregation of monetary measures in accounting discussed thus far can be examined in a two-dimensional (debit-credit) Cartesian space with two coordinates, one for the debit and one for the credit. This space is two-dimensional since the aggregation of monetary measures at each level in accounting is performed with respect to the debits and credits of individual accounts or groups of accounts. In the two-dimensional setting, every monetary measure, whether primitive or derived, or debit or credit, is considered a point or line on the debit or credit coordinate. Aggregation at each level is performed by computing the combined distance of those multiple collinear points from the point of origin (zero monetary quantity) or the combined length of those multiple collinear lines, along each coordinate.

Level 1 Aggregation The level 1 aggregation of accounting measures is performed to get the total debits and total credits respectively of individual accounts at the end of each accounting period. To do so geometrically for each account, a debit-credit coordinate system is set up when the first accounting event involving the account is initially recorded during the accounting period. This is done just as a T-account for each account is set up in a ledger book. And each primitive measure (debit or credit) of the account is then presented as a point or line on the debit or credit coordinate. As subsequent accounting events involving the same account continue to take place during the accounting period, the primitive debits and credits of the account continue to be shown as points or lines on the same debit or credit coordinate.

When it comes to the end of the accounting period, for each account, there is a set of collinear points or lines on each side, positive and negative, of the debit and credit coordinates, respectively. Those points or lines on each coordinate represent all debits or all credits of the account. These debits and credits are primitive measures. The level 1 aggregation of the debits and the credits respectively of the account is then performed by computing the distance of those points combined or the length of those lines combined along the positive and negative sides respectively of the debit and credit coordinates. This is shown below.

$$|D_T| = \sum |d_k| \quad (k = 1, 2, \ldots, n)$$
$$|C_T| = \sum |c_k| \quad (k = 1, 2, \ldots, n)$$

where d_k and c_k are the point (line) for the kth debit amount and k^{th} credit amount respectively of the account in question on the respective debit and credit coordinates. The absolute value bars "| |" denote a "distance" or "length." D_T stands for the total debits of the account, and C_T for the total credits. It is also a point or line on the same debit or credit coordinate. The points d_k and c_k each are shown as a negative number if they mean a decrease in the account (e.g., asset or liability). The negative number, however, does not mean that the decrease in monetary quantity is intrinsically

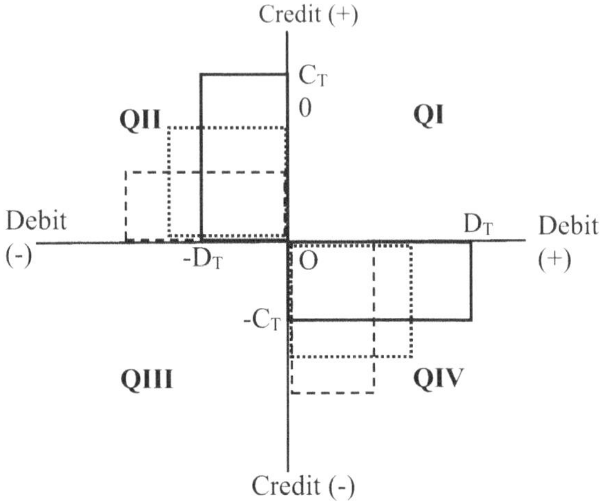

Fig. 8.1 The level 1 aggregates of individual accounts in the debit-credit geometric space

negative. Showing the decrease as a negative number is a matter of presentation, not substance.

For each account, the level 1 aggregation yields one of the three possible results:

$$|D_T| >, =, \text{or} < |C_T|$$

That is, the distance of the point D_T from the point of origin is longer or shorter than, or equal to, the distance of the point C_T.

Figure 8.1 graphically shows the possible results of level 1 aggregation for different accounts. They are shown by two rectangles and one square in the second and fourth quadrants (QII and QIV) respectively of the debit-credit Cartesian space. Each of them is defined by a pair of level 1 aggregates, D_T and C_T, of an account. The results of aggregation shown in QIV concern asset and expense accounts, and those shown in QII concern liability, equity, and revenue accounts. The result of level 1 aggregation shown by the solid-line rectangle in each quadrant is more typical than the result shown by the dashed-line rectangle or the dotted-line square. That is, it is more typical that $D_T > C_T$ for asset and expense accounts and that $C_T > D_T$ for liability, equity, and revenue accounts. The total credits (total decreases) of asset accounts are shown as $-C_T$, not $-D_T$, and the total debits (total decreases) of liability and equity accounts are shown as $-D_T$, not $-C_T$. This is because of the way the debits and credits of those accounts as decreases are shown in the journal entries. For all accounts, whether balance sheet or income statement, QI and QIII are not relevant to their level 1 aggregations because no account can have both D_T and C_T as total increases and both $-D_T$ and $-C_T$ as total decreases.

The solid-line rectangle in QIV represents the result of level 1 aggregation for asset accounts with $|D_T| > |-C_T|$, that is, $D_T > C_T$. The point D_T or line OD_T represents total debits (increases) of an asset account, and the point $-C_T$ or the related line represents the total credits (decreases) of the asset account. The point D_T is determined by the combined distance of the points d_k's (primitive debits) on the positive side of the debit coordinate (although not shown in the figure). The point $-C_T$ is determined similarly, by the combined distance of the points c_k's (primitive credits) on the negative side of the credit coordinate (not shown in the figure). The solid-line rectangle in QII represents the result of level 1 aggregation for liability and equity accounts with $|C_T| > |-D_T|$, that is, $C_T > D_T$. The point C_T or line OC_T represents total credits (increases) of a liability, or an equity, account, and the point $-D_T$ or the related line represents the total debits (decreases) of the same account. The point C_T is determined as is D_T for the asset account above, except that it is determined by the combined distance of the points c_k's on the positive side of the credit coordinate. The point $-D_T$ is also determined as is the point $-C_T$ for the asset account above, except that it is done with $-d_k$'s on the negative side of the debit coordinate. The dashed-line rectangle in each quadrant means the opposite of what the solid-line rectangle in the same quadrant means. The dotted-line square in each quadrant shows the result of level 1 aggregation: $|D_T| = |-C_T|$ or $|C_T| = |-D_T|$. For most accounts, either of these cases is likely rare as indicated earlier.

As for revenue accounts, the solid-line rectangle in QII is more relevant than the dashed-line rectangle or the dotted-line square since their total debits cannot exceed or be equal to their total credits. It is rare that $D_T > C_T$ or $D_T = C_T$ for revenue accounts. In fact, even the solid-line rectangle degenerates into a point or line on the positive side of the credit coordinate since $d_k = 0$ and hence $D_T = 0$ unless there is an adjustment or a correction on the debit side of the revenue accounts. For expense accounts, the solid-line rectangle in QIV is more relevant since their total credits cannot exceed or be equal to their total debits. Even this rectangle degenerates into a point or line on the positive side of the debit coordinate, since $c_k = 0$ and hence $C_T = 0$ unless there is an adjustment or a correction on the credit side of the expense accounts.

While QII and QIV of the debit-credit space are relevant to the presentation of the results of level 1 aggregation for individual accounts, QI and QIII are not. This is because no account, whether asset or liability or owner equity, can have both d_k and c_k as either an increase or a decrease. Nor can any account have both D_T and C_T as total increases and both $-D_T$ and $-C_T$ as total decreases.

Level 2 Aggregation Showing geometrically how the level 2 aggregation of monetary measures works is not as easy since it is performed in two steps. In the first step, C_T for each account is offset against D_T, or vice versa, in order to get the net change D_C or C_C for the account. Figure 8.2 shows how the first step of level 2 aggregation is performed geometrically. For asset accounts with $D_T > C_T$ (i.e., $|D_T| > |-C_T|$), a new point (D_T') is set on the positive side of the debit coordinate such that $|D_T'| = |-C_T|$ in distance from the point of origin. This is shown by the curved dashed line in QIV. The net change D_C is then determined as: $D_C = D_T - D_T'$. It is determined this way since $-C_T$ and D_T do not lie on the same debit coordinate. For liability and equity accounts, the net change C_C is also determined similarly, except that it is done in QII on the positive side of the credit coordinate. For income statement accounts, the first step of level 2 aggregation is performed with $D_T = 0$ for revenue accounts and $C_T = 0$ for expense accounts. Or it

Fig. 8.2 Determining the net changes in individual accounts for level 2 aggregation in the debit-credit geometric space

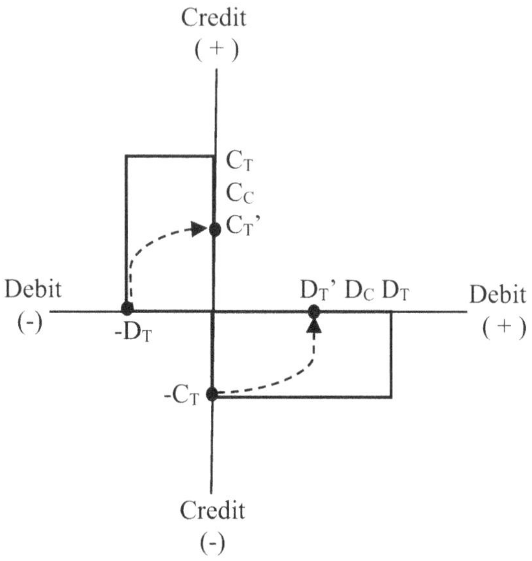

is skipped since $C_C = C_T$ for revenue accounts and $D_C = D_T$ for expense accounts as indicated before. Revenue accounts cannot have a debit entry as a primitive measure since there is no such thing as a negative or zero revenue. Thus, $d_k = 0$ and $D_T = 0$. Similarly, expense accounts cannot have a credit entry as a primitive measure since there is no such thing as a negative or zero expense. Thus, $c_k = 0$ and $C_T = 0$. This is so unless there is an adjustment or a correction of those accounts.

The second step of level 2 aggregation for individual accounts is performed in order to get their period-end balance E_B. Geometrically, it is done by adding the distance of the point D_C or C_C and the distance of the point B_B—the period-beginning balance of each account—along the positive side of each coordinate.[1] For balance sheet accounts, the point B_B was set on the positive side of the same debit or credit coordinate at the end of the previous accounting period. For income statement accounts, $B_B = 0$ since they are closed at the end of each accounting period. Thus, the second step of level 2 aggregation is performed with $D_T = 0$ for revenue accounts and $C_T = 0$ for expense accounts on the positive side of the debit and credit coordinates, respectively. Or it can be skipped. Thus, $E_B = C_C = C_T$ and $|E_B| = |C_C| = |C_T|$ for revenue accounts, and $E_B = D_C = D_T$ and $|E_B| = |D_C| = |D_T|$ for expense accounts.

Figure 8.3 shows the results of level 2 aggregation for different accounts. The level 2 aggregates are shown by the point E_B on the positive side of each coordinate. For asset accounts, this point is equal to $|B_B| + |D_C|$ in distance. For liability and equity accounts, it is equal to $|B_B| + |C_C|$ in distance. For revenue

[1] In the figure, the point B_B is not shown (to keep the figure simple). Rather, it is assumed to be the same point as D_T' or C_T' on the positive side of the respective coordinate (for convenience).

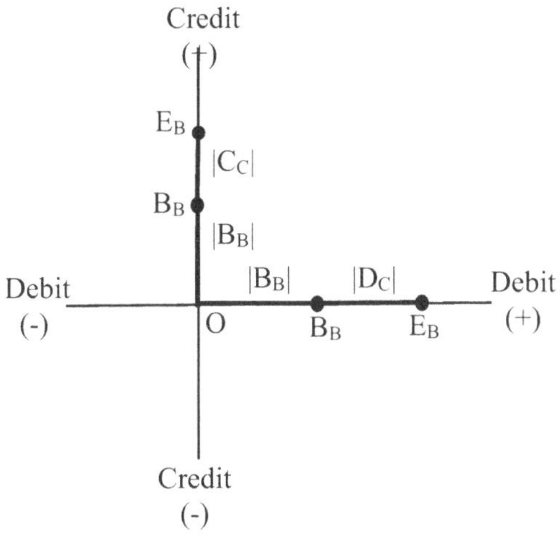

Fig. 8.3 The results of level 2 aggregation for individual accounts in the debit-credit geometric space

accounts, $|E_B| = |C_C| = |C_T|$ as $B_B = 0$ and the point E_B lies on positive side of the credit coordinate. For expense accounts, $|E_B| = |D_C| = |D_T|$ as $B_B = 0$ and the point E_B lies on the positive side of the debit coordinate.

Level 3 Aggregation The level 3 aggregation of monetary measures is performed by adding the level 2 aggregates (E_B) of similar accounts as a group, in order to get group totals such as current or total assets, current or total liabilities, total revenues, and total expenses. To do so geometrically, a separate debit-credit coordinate system is set up for each group of accounts at the end of the accounting period. To get a level 3 aggregate of asset accounts (say, total assets), the level 2 aggregates (E_B) of individual asset accounts in the group are shown as a set of points on the positive side of the debit coordinate. The combined distance of these points from the point of origin is then computed, which determines the point for the level 3 aggregate (total assets) of the asset group on the positive side of the debit coordinate. The point for the level 3 aggregate of liability accounts (say, total liabilities) is determined similarly, except that it is determined on the positive side of the credit coordinate. The point for the level 3 aggregate of owner equity accounts is also determined in the same way.

The point (or line) for the level 3 aggregate of income statement accounts is also determined similarly. To compute total revenues and total expenses, respectively, a separate debit-credit coordinate system is set up at the end of the accounting period, just as an income summary account is created at the end of each accounting period. For revenue accounts, their level 2 aggregates (E_B) are shown as a set of points or lines on the positive side of the credit coordinate. The point for their level 3 aggregate (total revenues) is then determined by the distance of those points combined or the length of those lines combined. For expense accounts, the point (line) for their level 3 aggregate (total expenses) is determined similarly, except that it is done along the positive side of the debit coordinate. The result is a pair of points, one for total revenues and one for total expenses, on the positive side of the debit and credit coordinates, respectively.

Given the total revenues and total expenses, the level 3 aggregation process for income statement accounts continues on, in order to determine the net income of the accounting period and include it in the retained earnings account. To do so geometrically, a point for total revenues (R_T) and total expenses (E_T) respectively is set on the positive side of the debit and credit coordinates, respectively. A new point (R_T') representing total expenses is set on the positive side of the credit coordinate such that $|R_T'| = |E_T|$ in distance from the point of origin. This is done in similar ways the point D_T' is set to determine D_C, and C_T' is set to determine C_C, as part of the level 2 aggregation for balance sheet accounts before. The net income is then determined to be equal to the distance between R_T' and R_T (or the length of the line between the two points) on the positive side of the credit coordinate. Once determined, it is considered a part of the net change (C_C) in the retained earnings account and included in the period-end balance E_B of the account. This balance is then included in setting the point (line) of total owner equity, which is an ultimate level 3 aggregate.

The level 3 aggregates as group totals are the ultimate aggregates of all accounts in use by the business entity (firm) in question. As such, they are not (or cannot be) aggregated any further. They are total assets, total liabilities, and total equity of the firm. They are positive as no firm can have negative total assets or, more generally, negative economic (capital) resources.[2] Once they are determined, the following holds: total assets = total liabilities + total equity, which is known as the fundamental accounting model. This model is a mathematical presentation of the ultimate level 3 aggregates of all accounts in the books of accounts of the firm.

The shaded square in Fig. 8.4 shows the ultimate level 3 aggregates of all accounts as group totals and their relationship as shown by the fundamental accounting model in the debit-credit geometric space. The point A denotes total assets and S denotes total liabilities plus total owner equity. Conceptually, the shaded square is considered to reflect the duality and balance features of the capital resources of the firm in question. The diagonal (line OD) of the square shows what the shaded square means. It has an angle of 45° with the coordinates and a slope of 1. Thus, it reflects the same duality and balance features of the capital resources of the firm as the model does. Having a slope of 1, it shows that the relationship between the dual sides (source and use) of the capital resources is a perfectly balanced one.

[2] A firm may have a negative owner equity on books. This, however, does not mean that the monetary value of the equity of the firm is intrinsically negative. A firm with a negative owner equity, however, is not considered normal. If the owner equity continues to be negative, the firm is unlikely to exist as a "going-concern." Conceptually, the existence of such firms is not consistent with the going-concern assumption in accounting.

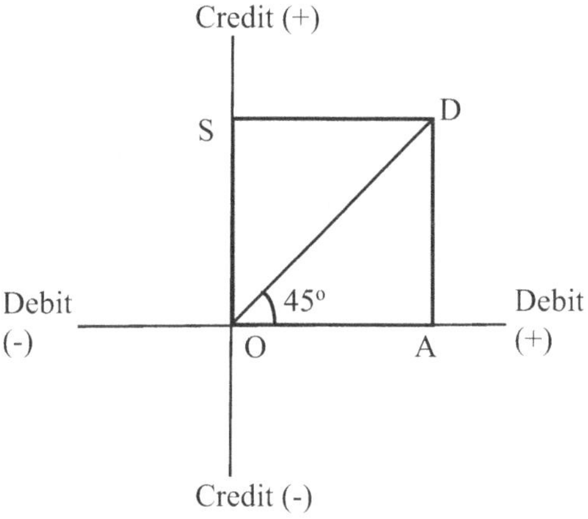

Fig. 8.4 The ultimate level 3 aggregates and their relationship in the debit-credit geometric space

8.4 Aggregating the Measures of Adjunct Accounts

Some accounts, whether they are balance sheet accounts or income statement accounts, are accompanied by the so-called *adjunct* accounts. Examples include accumulated depreciation, purchase discount, sales discount, bond premium or discount, and treasury stock, among others. They are called adjunct accounts since they cannot exist on their own but exist only in association with their main accounts. They are also called valuation accounts in that they need to be considered in determining the period-end balance E_B of their main accounts.

For adjunct accounts, the level 1 and level 2 aggregations of their monetary measures (debits and credits) are performed as usual, namely in the same way the level 1 and level 2 aggregations of the monetary measures of their main accounts are. Once their level 2 aggregate (period-end balance E_B) is determined, it is added to, or deducted from, the period-end balance (E_B) of their main accounts. Whether it is added or deducted depends on the nature of the adjunct account; it is deducted if the adjunct account is the so-called a *contra* account such as purchase discount, sales discount, and accumulated depreciation, for example. A contra account changes in the opposite direction of its main account so that its level 2 aggregate is considered a decrease in the balance of the main account. Hence, the level 2 aggregate of the contra account is deducted when the level 2 aggregate of its main account is determined. The level 2 aggregates of other adjunct accounts are added to the level 2 aggregates of their main accounts since they change in the same direction of their main accounts. Examples of such adjunct accounts include a bond premium and an additional paid-in capital in excess of the par value of capital stock.

For individual (main) accounts having an adjunct account, their period-end balance (E_B) is determined as follows:

$$\begin{aligned} E_B &= B_B \pm D_C \pm v \\ &= B_B \pm C_C \pm v \end{aligned}$$

where v is the level 2 aggregate (E_B) of the valuation (adjunct) account. It is subtracted to determine E_B for the main account if the valuation account is a contra account. Otherwise, it is added. The first aggregation above concerns asset and expense accounts, and the second one concerns liability, owner equity, and revenue accounts.

The level 2 aggregation of an account having an adjunct account can be explained geometrically. First, the point (or line) v, which is the level 2 aggregate (E_B) of an adjunct account, is set on the debit or credit coordinate on which E_B for the main account is shown. The distance of the point (or length of the line) is then added to, or subtracted from, the distance of the point E_B (or length of the line) for the main account. It is deducted if the adjunct account is a contra account; otherwise, it is added. This updates the point E_B for the main account and re-sets its location on the same debit or credit coordinate. If the adjunct account is a contra account, the point v does not lie on the same positive side of the debit or credit coordinate on which E_B for the main account is shown. Thus, a new point for v is set on the same side of the debit or credit coordinate in the same way the point D_T' or C_T' is set on the positive side of the debit or credit coordinate as D_C and C_C were determined in Fig. 8.2. Once this is done, the distance of the new point is deducted and the point E_B for the main account is updated. To determine total *net* sales as a level 3 aggregate, for example, a new point for v as the level 2 aggregate of the sales discounts account is set on the positive side of the credit coordinate on which the point E_B for total sales revenue is shown. The level 3 aggregate of the sales revenue account is then computed by deducting the distance of the new point for v from the distance of E_B for the sales revenue account. This updates E_B for the revenue account and re-sets its location on the positive side of the credit coordinate.

As for the adjunct accounts themselves, the level 3 aggregation is not performed since, unlike their main accounts, they are not aggregated or reported individually or as a group on the financial statements. For example, sales discounts are not aggregated across different sales revenue accounts in order to report their total as a level 3 aggregate on the income statement. Nor are purchase discounts aggregated across different purchase accounts for the same purpose. The same is also true for bond discounts or premiums, for example. They are not aggregated across different bonds-payable accounts in order to report their total on the balance sheet.

8.5 Aggregating Accounting Measures in Vector Form

The aggregations of monetary measures in accounting, regardless of their levels, can be examined in the two-dimensional debit-credit vector space. In a vector space, a point is a vector, and a line is a pair of vectors—one for each end of the line. In a vector space, therefore, all monetary measures, whether primitive or derived, of individual accounts can be shown as points or lines, hence vectors. These vectors are not any vectors but coordinate vectors. They lie on, not off, the debit and credit coordinates. The aggregation of monetary measures for individual accounts at each level is performed in the form of vector addition along the respective debit and credit coordinates. The vectors being added should be collinear and conformable for addition, meaning that they point in the same direction and have the same order—the same number of elements.

Mathematically, vectors can be added or subtracted, regardless of whether they are collinear or not, as long as they are conformable for addition (subtraction). And the resultant sum (or difference) vector has a meaning. This, however, is not true for the aggregation of vectors for monetary measures in accounting. Since the vectors of monetary measures have a direction, their addition or subtraction does not always yield a vector for a meaningful aggregate, unless they are collinear. If, for example, the vector for the total credits of an asset account on the credit coordinate is subtracted from the vector for the total debits of the same asset account on the debit coordinate, the resultant difference vector does not represent the net change in the asset. Similarly, if the vector for the total debits of a liability account on the debit coordinate is subtracted from the vector for the total credits of the liability account on the credit coordinate, the resultant difference vector does not represent the net change in the liability. To add or subtract the monetary measures of individual accounts in vector form, the vectors of those measures should be not only conformable for addition (or subtraction) but also collinear. And they should lie on the same (debit or credit) side of the same coordinate. Otherwise, the resultant sum or difference vector has no meaningful interpretation as an aggregate in accounting. Aggregating monetary measures in vector form means reducing a set of collinear vectors to a single vector along the same (positive or negative) side of the debit and credit coordinates, respectively.

Besides meeting the collinearity requirement above, the vectors representing monetary measures should also be independent even though they are collinear, regardless of whether they lie on the debit or credit coordinate. In other words, they represent the monetary measures that come from different accounting events or relate to different accounts. To get the vector for a level 1 aggregate, total debits or total credits, of each account, the vectors of the primitive debits and credits respectively of the account should come from unrelated accounting events. Similarly, to get the vector for a level 3 aggregate, the vectors of the level 2 aggregates being added should relate to different accounts in the same group. That is, the collinear vectors being added represent independent monetary measures, whether these measures are primitive or derived, even if those vectors are collinear and have the same magnitude. The independence-of-measures requirement, also discussed in Sect. 8.1, is important since the aggregation of monetary measures, whether debits or credits, in vector form is performed as a linear combination of vectors for monetary measures, each scaled by a factor of 1, not just any scalar.

Level 1 Aggregation The primitive measures, debits or credits, of each account are points or lines on the respective debit and credit coordinates. These points or lines are coordinate vectors of order 2 in the two-dimensional debit-credit vector space. The vectors on each side, debit or credit, of the respective coordinates are collinear yet independent since they represent the debits and credits that originate from different accounting events. Thus, they can be linearly added to get a new vector as shown below.

$$\begin{aligned}\mathbf{d}_T &= \sum \mathbf{d}_k = (d_1, 0) + (d_2, 0) + \ldots + (d_n, 0) = (D_T, 0)\\ \mathbf{c}_T &= \sum \mathbf{c}_k = (0, c_1) + (0, c_2) + \ldots + (0, c_n) = (0, C_T)\end{aligned}$$

The new vector $\mathbf{d}_T$ represents the total debits of the account and $\mathbf{c}_T$ the total credits. As a pair, $\mathbf{d}_T$ and $\mathbf{c}_T$ represent the pair of level 1 aggregates of the account. The component vector $\mathbf{d}_k$ ($\mathbf{c}_k$) represents the kth primitive debit (credit) of the account. Both $\mathbf{d}_k$ and $\mathbf{c}_k$ are coordinate vectors of order 2. They are component vectors, and $\mathbf{d}_T$ and $\mathbf{c}_T$ each are a linear combination of the related component vectors. The component debit vectors $\mathbf{d}_1$, $\mathbf{d}_2$, ... , and $\mathbf{d}_n$ are collinear, independent, and conformable for addition so that they can be linearly added to get $\mathbf{d}_T$ as a sum vector. The same is also true for $\mathbf{c}_1$, $\mathbf{c}_2$, ... , and $\mathbf{c}_n$. Thus, these component vectors can be linearly added to get $\mathbf{c}_T$ as a sum vector. The component vectors, whether debit or credit, are mutually independent since they represent the primitive measures that come from unrelated accounting events. For revenue accounts, $\mathbf{d}_T$ is a null vector as $\mathbf{d}_k = \mathbf{0}$ and, for expense accounts, $\mathbf{c}_T$ is a null vector as $\mathbf{c}_k = \mathbf{0}$, unless there is an adjustment or a correction.[3] Generating $\mathbf{d}_T$ and $\mathbf{c}_T$ as a linear combination of the component vectors is consistent with the concept of linear aggregation and the "whole-parts" relation mentioned before.

For each account, $\mathbf{d}_T$ and $\mathbf{c}_T$ are coordinate vectors and mutually orthogonal. But they may or may not have the same magnitude. Therefore, the following hold for the results of the level 1 aggregation of the debits and credits respectively of the account:

$$|\mathbf{d}_T| >, =, \text{or} < |\mathbf{c}_T| = |\mathbf{a}_T|$$

and

$$|\mathbf{c}_T| >, =, \text{or} < |\mathbf{d}_T| = |\mathbf{b}_T|$$

The first expression concerns asset and expense accounts, and the second expression concerns liability, equity, and revenue accounts. The vector $\mathbf{c}_T$ for asset and expense accounts means total credits (total decreases), hence shown by $\mathbf{a}_T$. And $\mathbf{d}_T$ for liability, equity, and revenue accounts means total debits (total decreases), hence shown by $\mathbf{b}_T$.[4] For revenue accounts, $|\mathbf{d}_T| = |\mathbf{b}_T| = 0$ and $|\mathbf{c}_T| > 0$) and, for expense accounts, $|\mathbf{c}_T| = |\mathbf{a}_T| = 0$ and $|\mathbf{d}_T| > 0$, unless there is an adjustment or a correction. For asset accounts, $\mathbf{a}_T$ represents total credits (decreases) so that it lies on the

[3]The same is also true for $\mathbf{d}_T$ or $\mathbf{c}_T$ for the adjunct accounts discussed before.

[4]These vector notations are consistent with those in Chap. 6.

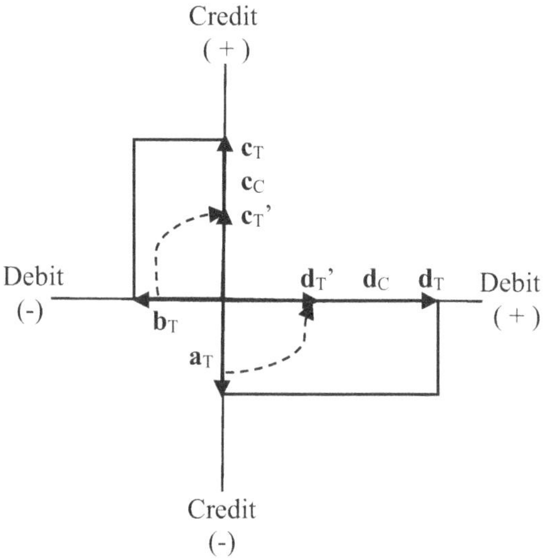

Fig. 8.5 The level 1 aggregates of individual accounts and their net changes for level 2 aggregation in the debit-credit vector space

negative side of the credit coordinate. For liability and owner equity accounts, $\mathbf{b}_T$ represents total debits (decreases) so that it lies on the negative side of the debit coordinate. As a pair of adjacent coordinate vectors, $\mathbf{d}_T$ and $\mathbf{a}_T$ are mutually orthogonal. So are $\mathbf{c}_T$ and $\mathbf{b}_T$.

Figure 8.5, adapted from Fig. 8.1, graphically shows a typical result of level 1 aggregation for different kinds of accounts in the two-dimensional vector space. For asset accounts, $\mathbf{d}_T$ is a debit coordinate vector and $\mathbf{a}_T$ is a credit coordinate vector. And the magnitude of $\mathbf{d}_T$ is assumed to be larger than the magnitude of $\mathbf{a}_T$ ($D_T > C_T$). Together, the pair of coordinate vectors defines the rectangle in QIV of the debit-credit vector space. Similarly, for liability and owner equity accounts, $\mathbf{c}_T$ is a credit coordinate vector and $\mathbf{b}_T$ is a debit coordinate vector. And the magnitude of $\mathbf{c}_T$ is assumed to be larger than the magnitude of $\mathbf{b}_T$ ($C_T > D_T$). Together, they define the rectangle in QII. For revenue accounts, $D_T = 0$ so that the rectangle in QII degenerates into a point or line, thus the vector $\mathbf{c}_T$, on the positive side of the credit coordinate (unless there is an adjustment or a correction). For expense accounts, $C_T = 0$ so that the rectangle in QIV degenerates into a point or line, thus the vector $\mathbf{d}_T$, on the positive side of the debit coordinate. For most accounts, whether balance sheet or income statement, it is rare that $|\mathbf{d}_T| = |\mathbf{a}_T|$ and $|\mathbf{c}_T| = |\mathbf{b}_T|$ as indicated before. Thus, these cases are not shown in the figure.

Level 2 Aggregation The level 2 aggregation of the debits and credits respectively of individual accounts in vector form is performed in two steps, the first step as vector subtraction and the second step as vector addition. For each account, the first step of level 2 aggregation yields the difference vector $\mathbf{d}_C$ or $\mathbf{c}_C$ for the net change in the account during the accounting period. This is shown below.

$$\mathbf{d}_C = \mathbf{d}_T - \mathbf{d}_T' = (D_C, 0)$$

or

$$\mathbf{c}_C = \mathbf{c}_T - \mathbf{c}_T' = (0, C_C)$$

where $\mathbf{d}_T$' and $\mathbf{c}_T$' are a new vector, which is created to perform the first step of level 2 aggregation. Every account has $\mathbf{d}_C$ or $\mathbf{c}_C$, not both. $\mathbf{d}_T$' is created on the debit coordinate and replaces $\mathbf{a}_T$ for the total credits (decreases) of asset and expense accounts. It is set on the positive side of the debit coordinate, in such a way that $|\mathbf{d}_T'| = |\mathbf{a}_T|$ in magnitude. It is collinear with, hence subtracted from, $\mathbf{d}_T$ on the positive side of the debit coordinate. Similarly, $\mathbf{c}_T$' is created on the credit coordinate and replaces $\mathbf{b}_T$ for the total debits (decreases) of liability, equity, and revenue accounts in similar ways, except that it is done on the positive side of the credit coordinate. The creation of the new vectors as described above is necessary to get $\mathbf{d}_C$ or $\mathbf{c}_C$ for the net change in each account. This is because $\mathbf{d}_T$ and $\mathbf{a}_T$ are not collinear so that $\mathbf{a}_T$ cannot be subtracted from $\mathbf{d}_T$ in order to get $\mathbf{d}_C$. Likewise, $\mathbf{b}_T$ cannot be subtracted from $\mathbf{c}_T$ in order to get $\mathbf{c}_C$.

Figure 8.5 shows how the new vectors $\mathbf{d}_T$' and $\mathbf{c}_T$' for two different accounts (e.g., asset and liability) are set on the positive side of each coordinate. They are set as indicated by the curved dashed line, in similar ways the corresponding points were set to determine D_C or C_C in Fig. 8.2. $\mathbf{d}_T$' concerns an asset account with total debits larger than total credits ($|\mathbf{d}_T| > |\mathbf{a}_T|$). And $\mathbf{c}_T$' concerns a liability (and an owner equity) account with total credits larger than total debits ($|\mathbf{c}_T| > |\mathbf{b}_T|$). Once set as shown, the new vectors can be subtracted from $\mathbf{d}_T$ and $\mathbf{c}_T$ respectively on the same side of the respective coordinates in order to get the difference vector $\mathbf{d}_C$ and $\mathbf{c}_C$, respectively.

For each account, $\mathbf{d}_C$ or $\mathbf{c}_C$ is collinear with another vector—$\mathbf{B}_B$ for the period-beginning balance of the same account. $\mathbf{B}_B$ was generated at the end of the previous accounting period. It is also a coordinate vector, collinear with $\mathbf{d}_C$ or $\mathbf{c}_C$, and lies on the positive side of the same debit or credit coordinate, depending on the nature of the account (e.g., asset or liability).

In the second step of level 2 aggregation for each account, $\mathbf{d}_C$ or $\mathbf{c}_C$ is added to (or subtracted from) $\mathbf{B}_B$, in order to get a new coordinate vector—$\mathbf{E}_B$ for the period-end balance of the account. Since $\mathbf{d}_C$ and $\mathbf{B}_B$, and $\mathbf{c}_C$ and $\mathbf{B}_B$, are collinear and lie on the same side of the same debit or credit coordinate, they can be linearly added (or subtracted) to get $\mathbf{E}_B$ as a new vector:

$$\mathbf{E}_B = \mathbf{B}_B \pm \mathbf{d}_C$$

or

$$= \mathbf{B}_B \pm \mathbf{c}_C$$

To get $\mathbf{E}_B$ for some accounts, $\mathbf{d}_C$ is subtracted if it represents D_C as a net decrease ($D_C < 0$). The same is also true for $\mathbf{c}_C$ if $C_C < 0$. The vector $\mathbf{E}_B$ in the first expression

Fig. 8.6 The results of level 2 aggregation for individual accounts in the debit-credit vector space

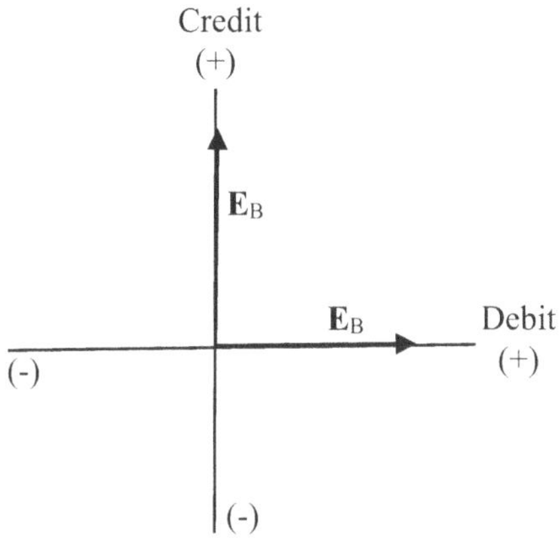

above concerns asset and expense accounts, and $\mathbf{E}_B$ in the second expression concerns liability, equity, and revenue accounts. For income statement accounts, $\mathbf{B}_B = \mathbf{0}$ since they are closed to compute the net income of the previous accounting period as indicated before. Thus, $\mathbf{E}_B = \mathbf{d}_C$ for expense accounts and $\mathbf{E}_B = \mathbf{c}_C$ for revenue accounts. For asset and expense accounts, $\mathbf{E}_B$ lies on the positive side of the debit coordinate. And, for liability, owner equity, and revenue accounts, $\mathbf{E}_B$ lies on the positive side of the credit coordinate.

Figure 8.6 presents the typical result of level 2 aggregation for individual accounts in the debit-credit vector space. $\mathbf{E}_B$ on the debit coordinate represents the period-end balance of asset and expense accounts, and $\mathbf{E}_B$ on the credit coordinate represents the period-end balance of liability, equity, and revenue accounts. Regardless, $\mathbf{E}_B$ is a positive coordinate vector as shown.

Level 3 Aggregation The level 3 aggregation of monetary measures in the two-dimensional vector space is performed to get vectors for various group totals by adding the vectors ($\mathbf{E}_B$) for the period-end balances of similar accounts in the same group. It is performed in the form of a linear combination of vectors, each scaled by a factor of 1. The result is a sum vector, which represents a group total such as current or total assets, current or total liabilities, total equity, total revenues, or total expenses.

Shown below are the vectors of the ultimate level 3 aggregates of all asset accounts, all liability accounts, and all owner equity accounts, respectively.

$$\begin{aligned} \mathbf{A} &= \sum\nolimits_j \mathbf{E}_{Bj} = (E_{B1}, 0) + \ldots + (E_{Bn}, 0) = (A, 0) \\ \mathbf{L} &= \sum\nolimits_j \mathbf{E}_{Bj} = (0, E_{B1}) + \ldots + (0, E_{Bn}) = (0, L) \\ \mathbf{E}_q &= \sum\nolimits_j \mathbf{E}_{Bj} = (0, E_{B1}) + \ldots + (0, E_{Bn}) = (0, E_q) \end{aligned}$$

where the subscript j denotes the jth account in each account group. $\mathbf{A}$, $\mathbf{L}$, and $\mathbf{E}_q$ are coordinate vectors of order 2, each generated as a linear combination of its

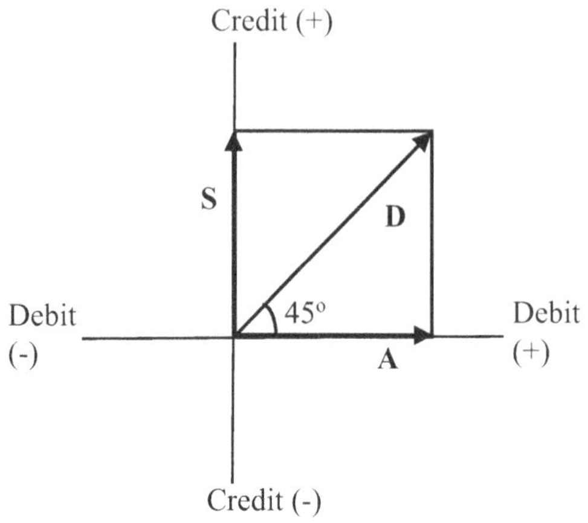

Fig. 8.7 The ultimate level 3 aggregates and their relationship in the debit-credit vector space

component vectors. The component vectors of **A**, **L**, and $\mathbf{E}_q$ respectively are collinear, independent, and conformable for vector addition so that they can be linearly added. Given **A**, **L**, and $\mathbf{E}_q$, the following holds in magnitude, which is what the fundamental accounting model shows:

$$|\mathbf{A}| = |\mathbf{S}|$$

where **S** is a sum vector defined as $\mathbf{S} = \mathbf{L} + \mathbf{E}_q = (0, L + E_q) = (0, S)$.

The shaded square in Fig. 8.7, adapted from Fig. 8.4, shows **A** and **S** and their relationship in the debit-credit vector space. The vector **A** on the debit coordinate represents total assets (A), namely the uses of the capital resources, of the firm in question, and **S** on the credit coordinate represents the sum of total liabilities and total owner equity, namely the sources of the capital resources. As coordinate vectors, **A** and **S** are mutually orthogonal and have the same magnitude. Together, they define the shaded square. This square reflects the duality and balance features of the capital resources of the firm. The diagonal vector **D** is a linear combination of **A** and **S** as shown below.

$$\mathbf{D} = \mathbf{A} + \mathbf{S} = (A, S) = (A, L + Eq)$$

As a sum vector of order 2, **D** is equivalent to the shaded square. It represents the same ultimate level 3 aggregates of assets, liabilities, and owner equity of the firm, and their relationships. It reflects the duality and balance features of the capital resources of the firm as the shaded square does. As such, it has a slope of 1 since it has an angle of 45° with the adjacent debit and credit coordinates. **A** and **S** co-exist and co-vary and do so with balance.

8.6 Aggregating Accounting Measures in Matrix Form

The aggregation of monetary measures, whether primitive or derived, in matrix form is performed in similar ways those measures are aggregated in vector form. It begins with generating the vectors of the primitive and derived measures of individual accounts and then arranges these vectors as matrices. And the aggregation process ends with generating a matrix for the ultimate level 3 aggregates of all accounts as different group totals. The aggregation of monetary measures at each level is performed in the form of matrix addition. The matrices being added are conformable for addition, meaning that they have the same order—the same number of rows and columns—and the same rank. They are 2×1 or 2×2 matrices, depending on the level of aggregation. Each column (or row) of the matrices is considered a vector that represents a measure, primitive or derived. The geometric settings for all levels of aggregation in matrix form are two-dimensional vector spaces with two coordinates, debit and credit, similar to those for various levels of aggregation in vector form before.

Level 1 Aggregation To aggregate the primitive debits and credits respectively of individual accounts in matrix form, the point or line for the kth debit entry and k^{th} credit entry respectively of each account is considered a vector, $\mathbf{d}_k$ or $\mathbf{c}_k$, as it was in the previous section. This vector is now considered a 2×1 (or 1×2) matrix, depending on how its elements are arranged, row- or column-wise. The vectors $\mathbf{d}_k$ and $\mathbf{c}_k$ are generated when an accounting event involving the account takes place and is recorded during the accounting period. When it comes to the end of the period, there is a set of such debit and credit coordinate vectors for each account as different accounting events involving the same account continue to take place and be recorded during the period. They are then added to get $\mathbf{d}_T$ for the total debits (D_T), and $\mathbf{c}_T$ for the total credits (C_T), of the account. This is shown below.

$$\mathbf{d}_k = \begin{bmatrix} d_k \\ 0 \end{bmatrix}$$

$$\mathbf{c}_k = \begin{bmatrix} c_k \\ 0 \end{bmatrix}$$

$$\mathbf{d}_T = \sum \mathbf{d}_k = \begin{bmatrix} D_T \\ 0 \end{bmatrix}$$

$$\mathbf{c}_T = \sum \mathbf{c}_k = \begin{bmatrix} 0 \\ C_T \end{bmatrix}$$

where $\mathbf{d}_k$ and $\mathbf{c}_k$ are a two-element coordinate vector, or a 2×1 matrix, for the kth debit amount and k^{th} credit amount respectively of the account. Similarly, $\mathbf{d}_T$ and $\mathbf{c}_T$ are a two-element coordinate vector or a 2×1 matrix.

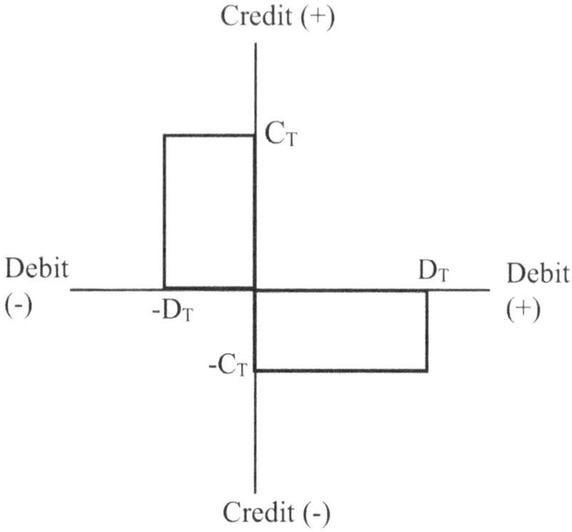

Fig. 8.8 The level 1 aggregates of individual accounts in matrix form in the debit-credit vector space

Once $\mathbf{d}_T$ and $\mathbf{c}_T$ are determined for each account, a 2 x 2 diagonal matrix ($\mathbf{A}_1$) is constructed as follows:

$$\mathbf{A}_1 = [\mathbf{d}_T \ \mathbf{c}_T] = \begin{bmatrix} D_T & 0 \\ 0 & C_T \end{bmatrix}$$

This matrix represents a pair of level 1 aggregates, total debits and total credits, of the account. The subscript 1 denotes a level 1 aggregation. Geometrically, $\mathbf{d}_T$ represents the point (D_T,0) on the debit coordinate, and $\mathbf{c}_T$ represents the point (0,C_T) on the credit coordinate. And D_T and C_T are the main diagonal elements of $\mathbf{A}_1$. The diagonal element D_T for revenue accounts, and C_T for expense accounts, is zero unless there is an adjustment or a correction of the accounts. Thus, for income statement accounts, $\mathbf{A}_1$ is written as

$$\mathbf{A}_1 = [\mathbf{d}_T \quad \mathbf{c}_T] = \begin{bmatrix} 0 & 0 \\ 0 & C_T \end{bmatrix} \text{or} \begin{bmatrix} D_T & 0 \\ 0 & 0 \end{bmatrix}$$

The first matrix above concerns revenue accounts and represents two points—the point of origin and the point C_T on the credit coordinate. And the second matrix concerns expense accounts and represents two points—the point of origin and the point D_T on the debit coordinate.

Figure 8.8, adapted from Fig. 8.5, graphically shows the possible results of level 1 aggregation for two different accounts. The shaded rectangle in QIV is a vector space represented by $\mathbf{A}_1$ for asset accounts with $|D_T| > |-C_T| > 0$. The point $-C_T$ represents total credits (decreases). It is shown by $-C_T$, not C_T, on the negative side of the credit coordinate since decreases in asset accounts are credited in journal entries. The shaded rectangle in QII is a vector space represented by $\mathbf{A}_1$ for liability and owner equity accounts with $|C_T| > |-D_T| > 0$. The point $-D_T$ represents total debits (decreases). It is shown by $-D_T$, not D_T, on the negative side of the debit coordinate since decreases in liability and equity accounts are debited in journal entries. For revenue accounts, the rectangle in QII degenerates into a line—a vector on the credit coordinate. For expense accounts, the rectangle in QIV degenerates into a line—a vector on the positive side of the debit coordinate. This is because $|-D_T| = 0$ for revenue accounts and $|-C_T| = 0$ for expense accounts, unless there is an adjustment or a correction of the accounts.

$\mathbf{A}_1$ is a 2 × 2 diagonal matrix or, more generally, a 2 × 2 symmetric matrix. As it is, $\mathbf{A}_1$ has general properties of diagonal or symmetric matrices. It equals its transpose: $\mathbf{A}_1 = \mathbf{A}_1{}^T$. It has a nonzero determinant ($|\mathbf{A}_1| \neq 0$), and this determinant is equal to the determinant of its transpose ($|\mathbf{A}_1| = |\mathbf{A}_1{}^T|$). Hence, it is nonsingular and has an inverse although this is not the case for income statement accounts. The inverse of $\mathbf{A}_1$ is equal to the transpose of its inverse: $\mathbf{A}_1{}^{-1} = (\mathbf{A}_1{}^{-1})^T$. And the product of $\mathbf{A}_1$ and its inverse yields an identity matrix: $\mathbf{A}\mathbf{A}_1{}^{-1} = \mathbf{I}$, where $\mathbf{I}$ is the identity matrix. Yet $\mathbf{A}_1$ has no relation with the duality and balance features of accounting events as it represents the total debits and total credits of individual accounts, not accounting events. Nor does it have any direct relation with the contrasting yet symmetric nature of double entries. Unlike the debit(s) and credit(s) of an accounting event, the total debits of an account do not need to match its total credits.

Geometrically, $\mathbf{A}_1$ for balance sheet accounts is considered to represent a vector space spanned by $\mathbf{d}_T$ and $\mathbf{c}_T$ as a pair of basis vectors. This vector space is represented by $\mathbf{A}_1$. It is a rectangle or square, depending on whether $|D_T| = |C_T|$ or not. For most balance sheet accounts, it is likely to be a rectangle such as the one in the figure since it is rare that $|D_T| = |-C_T|$ or $|C_T| = |-D_T|$. Regardless, the size of the vector space represented by $\mathbf{A}_1$ is equal to the determinant of $\mathbf{A}_1$ in absolute terms, which is computed as

$$|\mathbf{A}_1| = \begin{vmatrix} D_T & 0 \\ 0 & C_T \end{vmatrix} = |D_T.C_T - 0| = |D_T.C_T|$$

For income statement accounts, $|\mathbf{A}_1| = 0$. Thus, it is singular and has no inverse. The vector space represented by $\mathbf{A}_1$ is a null space, namely a line (or point) on the debit or credit coordinate, unless there is an adjustment or a correction of those accounts.

Level 2 Aggregation The level 2 aggregation in matrix form for each account begins by generating the vector $\mathbf{d}_C$ or $\mathbf{c}_C$ for the net change in the account during the accounting period. Both of these vectors are a two-element vector or are considered a 2 × 1 (or 1 × 2) matrix. The process of generating $\mathbf{d}_C$ and $\mathbf{c}_C$ was explained in the previous section as part of the level 2 aggregation in vector form. Every account has $\mathbf{d}_C$ or $\mathbf{c}_C$, not both, regardless of whether it is a balance sheet or income statement account.

Once $\mathbf{d}_C$ or $\mathbf{c}_C$ is obtained for each account, it is added to or subtracted from another 2 × 1 (or 1 × 2) matrix—$\mathbf{B}_B$ for the period-beginning balance (B_B) of the account. This is shown below.

$$\mathbf{B}_{\mathrm{B}} = \begin{bmatrix} \mathrm{B_B} \\ 0 \end{bmatrix} \text{or} \begin{bmatrix} 0 \\ \mathrm{B_B} \end{bmatrix}$$

$$\mathbf{A}_2 = \mathbf{B}_{\mathrm{B}} \pm \mathbf{d}_{\mathrm{C}} = \begin{bmatrix} \mathrm{B_B} \pm \mathrm{D_C} \\ 0 \end{bmatrix} = \begin{bmatrix} \mathrm{E_B} \\ 0 \end{bmatrix}$$

or

$$\mathbf{A}_2 = \mathbf{B}_{\mathrm{B}} \pm \mathbf{c}_{\mathrm{C}} = \begin{bmatrix} 0 \\ \mathrm{B_B} \pm \mathrm{C_C} \end{bmatrix} = \begin{bmatrix} 0 \\ \mathrm{E_B} \end{bmatrix}$$

As a 2 × 1 or 1 × 2 matrix, $\mathbf{A}_2$ represents the level 2 aggregate of the account. $\mathbf{A}_2$ in the first form above concerns asset and expense accounts with $\mathrm{E_B}$ as their positive period-end debit balance, and $\mathbf{A}_2$ in the second form concerns liability, equity, and revenue accounts with $\mathrm{E_B}$ as their positive period-end credit balance. For income statement accounts, $\mathrm{B_B} = 0$ since they are closed at the end of each accounting period. Thus, $\mathrm{E_B} = \mathrm{C_C} = \mathrm{C_T}$ for revenue accounts, and $\mathrm{E_B} = \mathrm{D_C} = \mathrm{D_T}$ or expense accounts, unless there is an adjustment or a correction of the accounts. Yet $\mathbf{A}_2$ for revenue and expense accounts is still considered a 2 × 1 or 1 × 2 matrix.

For each account, $\mathbf{A}_2$ differs from $\mathbf{A}_1$ before. $\mathbf{A}_2$ is a 2 × 1 or 1 × 2 matrix and represents the period-end balance of the account, while $\mathbf{A}_1$ is a 2 × 2 diagonal matrix and represents the total debits and total credits of the account. Thus, $\mathbf{A}_2$ has a rank of 1, while $\mathbf{A}_1$ has a rank of 2. And $\mathbf{A}_2$ has no determinant, while $\mathbf{A}_1$ (for balance sheet accounts) has a (nonzero) determinant. As it is, $\mathbf{A}_2$ represents a null space, namely a point ($\mathrm{E_B}$) or line, while $\mathbf{A}_1$ (for balance sheet accounts) represents a vector space.

Level 3 Aggregation The level 3 aggregation of the level 2 aggregates of similar accounts as a group in matrix form is performed by adding the matrices of the level 2 aggregates of those accounts. It gives three new matrices of order 2 × 1 or 1 × 2 as shown below. They represent the ultimate level 3 aggregates of all assets (A), all liabilities (L), and total owner equity (Eq). They are written as

$$\mathbf{A} = \sum \mathbf{A}_{2\mathrm{k}} = \begin{bmatrix} \sum \mathrm{E_{Bk}} \\ 0 \end{bmatrix} = \begin{bmatrix} \mathrm{A} \\ 0 \end{bmatrix} \quad \text{(k for asset account k)}$$

$$\mathbf{L} = \sum \mathbf{A}_{2\mathrm{k}} = \begin{bmatrix} 0 \\ \sum \mathrm{E_{Bk}} \end{bmatrix} = \begin{bmatrix} 0 \\ \mathrm{L} \end{bmatrix} \quad \text{(k for liability account k)}$$

$$\mathbf{E}_{\mathrm{q}} = \sum \mathbf{A}_{2\mathrm{k}} = \begin{bmatrix} 0 \\ \sum \mathrm{E_{Bk}} \end{bmatrix} = \begin{bmatrix} 0 \\ \mathrm{E_q} \end{bmatrix} \quad \text{(k for equity account k)}$$

As with $\mathbf{A}_2$ before, each of the matrices above is considered a 2 × 1 matrix. The component matrix $\mathbf{A}_{2\mathrm{k}}$ represents the level 2 aggregate of the kth account in the account group. Each matrix above is a linear addition of its component matrices,

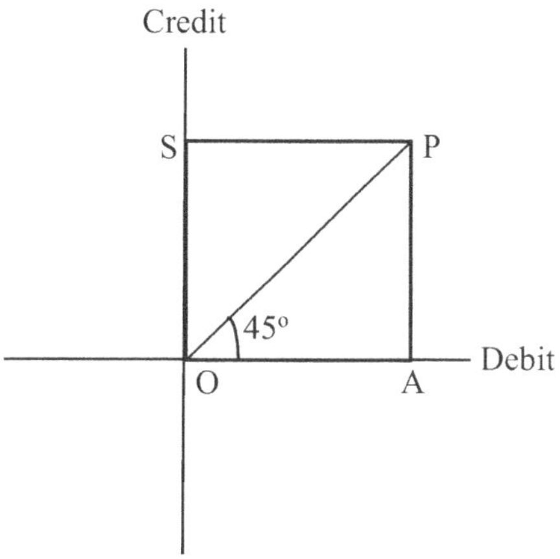

Fig. 8.9 The geometric space defined by the matrix of ultimate level 3 aggregates in the debit-credit vector space

each scaled by a factor of 1. Given **A**, **L**, and $\mathbf{E}_q$, another matrix ($\mathbf{A}_3$) is constructed as follows:

$$\mathbf{A}_3 = [\mathbf{A} \quad \mathbf{S}] = \begin{bmatrix} A & 0 \\ 0 & L + E_q \end{bmatrix} = \begin{bmatrix} A & 0 \\ 0 & S \end{bmatrix}$$

where $\mathbf{S} = \mathbf{L} + \mathbf{E}_q = (0, L + E_q) = (0, S)$. $\mathbf{A}_3$ is a 2 × 2 scalar matrix with a rank of 2, as its diagonal elements are the same scalar number. It represents the ultimate level 3 aggregates of all asset accounts and all liability and owner accounts respectively of a business firm. As a scalar matrix, $\mathbf{A}_3$ reflects the duality and balance features of the capital resources of the firm as the fundamental accounting model does.

Figure 8.9, adapted from Fig. 8.7, shows what $\mathbf{A}_3$ means geometrically. The point A on the debit coordinate represents the diagonal element A of $\mathbf{A}_3$, which represents total assets. And the point S on the credit coordinate represents the diagonal element S of $\mathbf{A}_3$, which represents the sum of total liabilities and owner equity. The shaded square is a vector space represented by $\mathbf{A}_3$. It is spanned by **A** and **S** as a pair of basis vectors and, as a square, reflects the duality and balance features of the capital resources of the business firm in question. Thus, the diagonal of the square has a slope of 1 as it has an angle of 45° with the debit and credit coordinates.

As a scalar matrix, $\mathbf{A}_3$ has a nonzero determinant, which is equal (in absolute value) to the size of the shaded square represented by $\mathbf{A}_3$: $|\mathbf{A}_3| = |A.S - 0| = A.S$. The diagonal elements A and S are positive ($A, S > 0$) since no firm can have zero or negative capital resources. Hence, the square represented by $\mathbf{A}_3$ lies in the first quadrant (QI) of the two-dimensional vector space. The diagonal (line OP), and the diagonal vertex (point P), of the square is equivalent to the square in that it represents what the square represents—the ultimate level 3 aggregates of all accounts and their relationship as shown by the fundamental accounting model. And it reflects the duality and balance features of the capital resources of the firm as the square does.

8.7 Comments on Aggregations in Accounting

Before closing the chapter, a few comments are in order regarding the aggregations of monetary measures in accounting. First, the aggregation of monetary measures at each level is performed with respect to individual *accounts* or a group of similar accounts, not accounting events. Aggregating the monetary (size) measures of accounting events gives no meaningful aggregate. Nor does it serve any purpose in accounting. The monetary quantities, debits or credits, of different accounting events are not homogeneous since they do not always involve the same account or similar accounts. Nor do they all mean increases or decreases in monetary quantity; rather, they can be a mix of increases and decreases and a mix of different accounts. As such, they are not homogeneous and cannot be linearly added to get a meaningful aggregate such as total debits or total credits, for example. To be meaningful, an aggregate should come from a set of homogeneous monetary measures—either debits or credits as increases or decreases, not a mix of both—and concern the same account or a group of similar accounts.

Second, unlike the case of real numbers, the aggregation of monetary measures in accounting always proceeds in one direction—from primitive measures to derived measures and from level 1 aggregation to level 3. Disaggregating aggregates in a reverse direction serves no purpose in accounting although it can be done mathematically. For example, disaggregating the total debits (or credits) of an account (e.g., cash or inventory) into its component primitive debits (credits), for whatever reason, is not a normal course of aggregation in accounting. Nor does it serve any purpose. Similarly, decomposing a level 3 aggregate as the "whole" into level 2 aggregates as its "parts" (i.e., total assets into individual assets) is not a part of the normal course of aggregations in accounting.

Third, the monetary measures being aggregated, whether primitive or derived, should not only be homogeneous in nature but also be measured in the same measurement unit (e.g., US dollar). If, for example, one primitive measure (debit or credit) of an account is measured in US dollar and another measure of the same account in euro, adding the two measures (debits or credits) to get the total debits or total credits of the account, or subtracting one from the other to get a change in the account, gives no meaningful aggregate. To get a meaningful aggregate, the measures being aggregated should be measured and shown in the same monetary unit. For multiplication or division, the things involved may be measured and shown in different measurement units, and the result can still have a meaningful interpretation. For example, the total weights of a group of people can be computed by multiplying the weight of each person measured in pounds and the size of the group measured by the number of the people. Unlike this, however, monetary measures generated using different measurement units (e.g., US dollar and euro) cannot be linearly aggregated to get total debits or total credits as a meaningful aggregate, regardless of whether those measures are primitive or derived.

Finally, in order to be meaningful, an aggregate in accounting should come from a set of independent monetary measures, whether primitive or derived, because aggregation is a linearly additive concept. This requirement, referred to as the independence-of-measures requirement earlier, means that the debits of each account as primitive measures for level l aggregation should come from different (unrelated) accounting events. The same is also true for the credits of each account as primitive measures. Similarly, the level 2 aggregates that are added to get a level 3 aggregate as a group total should come from different accounts in the same account group. The independence-of-measures requirement is important in order to exclude the impact of any interaction among the measures being aggregated. Any interaction among the measures can make the resultant aggregate differ from their linear sum. Conceptually, it is not consistent with the notion of linear aggregation. Nor is it consistent with the "whole-parts" relation mentioned before. In order for the "whole" to be the same as the sum of its "parts," there should be no interaction among the parts. What this means is like: $var(a + b) = var(a) + var(b)$, not $var(a + b) = var(a) + var(b) + 2cov(a,b)$, where "var" and "cov" stand for variance and covariance, respectively.

The Geometry of Financial Ratios

9

This chapter examines financial ratios in geometric settings. Financial ratios do not differ much from ratios in general, except that they are generated with financial numbers. Although generating and presenting financial ratios is not a task of accounting, this chapter examines the subject since the financial numbers used to generate them come from accounting. These financial numbers are the level 2 or level 3 aggregates of accounting measures discussed in Chap. 8. Also, financial ratios and accounting events have one thing in common—duality. The former have dividend and divisor as their dual sides, and the latter have debit and credit. A relevant geometric setting for examining financial ratios is a two-dimensional Cartesian space since every financial ratio is defined by a pair of financial numbers—one for its dividend (numerator) and one for its divisor (denominator). The financial ratios examined in this chapter are those called *accounting ratios*, defined only with financial numbers from accounting. The so-called market-based financial ratios such as the price-to-earnings (P/E) ratio and the price-to-book value (P/B) ratio are not discussed since not all the numbers used to generate them come from accounting.

9.1 Ratios in General and Geometry

The Concept and Nature of Ratios in General

A ratio is a number that represents the quantity of one thing measured relative to the quantity of another thing compared in the ratio. One of the things is represented by the dividend (numerator) of the ratio, and the other by the divisor (denominator). The two things compared can be tangible or intangible, and the measure of their quantity is shown by a number. The ratio is generated by dividing one of the two numbers by the other. Thus, a ratio involves two numbers and is relative, in nature. And generating a ratio means reducing two numbers to a single number. This number is a scalar number, positive or negative.

© The Author(s), under exclusive license to Springer Nature Switzerland AG 2025

B. T. Ro, *The Geometry of Accounting*, Springer Texts in Business and Economics,
https://doi.org/10.1007/978-3-031-88892-2_9

As scalar numbers, ratios do not have a unique measurement unit of their own. Yet, in order for a ratio to be meaningful, the quantity of the two things compared in the ratio should be measured and shown in the same measurement unit. Otherwise, the ratio has no meaningful interpretation. If the quantities of the two things compared are measured in different measurement units, the number representing their relationship is measured and shown by a rate, which is conceptually distinguished from a ratio. The exchange rate of two different monetary currencies is an example (e.g., US dollar versus euro). As the dividend and divisor of ratios are measured in the same measurement unit, generating ratios is similar to adding or aggregating measures. For addition (or subtraction), the quantities of things being added (or subtracted) should be measured in the same measurement unit. Otherwise, the resultant sum (or difference) has no meaningful interpretation. If, for example, the weight of a person is measured in pound and the weight of another person in kilogram, adding the two weight measures to get their total weights gives no meaningful result, since pound and kilogram are not the same measurement unit.

To compare two things in a ratio, they should have a common attribute, which can be measured and shown in the same measurement unit. If, for example, the two things compared in a ratio are two different cars (say, A and B), they should have a common attribute such as a monetary value, which can be measured and shown in the same monetary unit (e.g., US dollar). Once their monetary value is measured and shown in the same monetary unit, the two cars can be compared based on the ratio of their monetary values. If the ratio of the price of car A to the price of car B is 2, this means that the former is two times more expensive than the latter. This is what the ratio 2 means. If the two things compared in a ratio have no such common attribute, there is nothing for which they can be compared. Thus, no meaningful ratio can be defined between them. If, for example, a ratio is generated by comparing the height of a person with the weight of another person, this ratio has no meaningful interpretation since their height and weight are not a common attribute and, hence, cannot be measured in the same measurement unit.

Ratios in Geometric Space

The geometric setting where ratios can be examined is a two-dimensional Cartesian space with two coordinates, one for the divisor of ratios and the other for the dividend. In this space, the divisor and dividend respectively of a ratio are a point or line on the respective coordinate. And the ratio itself is also a point or line, or even a space. Unlike the former, however, this point or line lies *off*, not on, the divisor and dividend coordinates. As the two-dimensional Cartesian space has an origin, a point for a ratio is not just any point but a point with a direction from the point of origin. Similarly, a line for a ratio is not just any line but a sloped line. And the slope of this line is determined relative to the divisor (or dividend) coordinate. The point and line representing the dividend and divisor respectively of the ratio lie on, not off, the respective coordinate. As these coordinates are orthogonal, the geometric space representing the ratio is a rectangle or square, depending on whether the line for the dividend is longer or shorter than, or equal to, the line for the divisor.

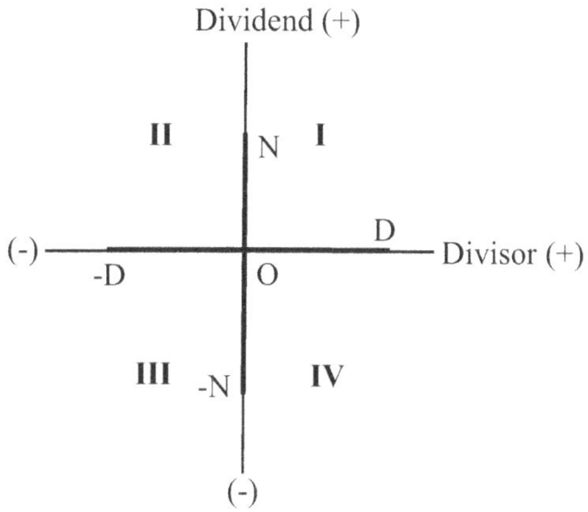

Fig. 9.1 A two-dimensional geometric setting for the presentation of ratios

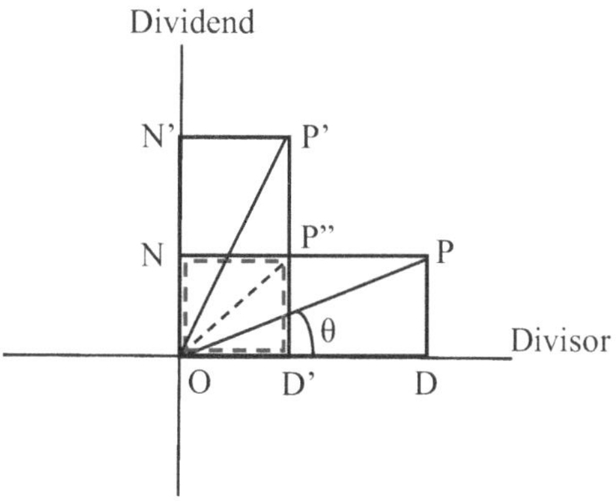

Fig. 9.2 Presenting three different types of ratios in the two-dimensional geometric space

Figure 9.1 presents an example of the two-dimensional Cartesian space where various types of ratios can be examined geometrically. The horizontal coordinate "Divisor" is reserved for the divisor (denominator D) of ratios, and the vertical coordinate "Dividend" for the dividend (numerator N). The point O, at which the divisor and dividend coordinates intersect, is the point of origin. Each of the two coordinates is divided into two sections, positive (+) and negative (−), based on the point O. This is because the divisor and dividend of ratios can be positive or negative numbers. The negative numbers are shown as −D and −N on the negative side of the respective coordinates. This, however, does not necessarily mean that the quantity of the thing represented by −D or −N is intrinsically negative. Whether it is shown by −D or −N, or D or −N, is a matter of presentation, not substance. The above specifications of the coordinate system divide the entire (divisor-dividend) Cartesian space into four quadrants, QI to QIV.

Figure 9.2 gives examples of three different types of ratios in terms of a point, line, and space respectively in the two-dimensional (divisor-dividend) space. The divisor and dividend of the ratios are assumed to be positive (D, N > 0 and D', N' > 0). Thus, the two rectangles (ODPN and OD'P'N') and the (dashed-line) square all lie in QI of the two-dimensional space. Each of them represents a distinct ratio. The related diagonal (line), and the diagonal vertex (point), also represents the same ratio. The ratio represented by the shaded rectangle is positive but less than 1 as $D > N$. If shown by a point, this ratio is determined by the direction, not distance, of the point P from the point of origin. If shown by a line, it is determined by the slope, not length, of the line OP—the diagonal of the rectangle—relative to the divisor coordinate. And, if shown by a space, it is determined by the shape, not size, of the shaded rectangle. The same is also true for the ratios represented by the corresponding point, line, and space relating to the unshaded rectangle and the dashed-line square, respectively, except that $D < N$ for the former and $D = N$ for the latter.

Of the alternative geometric presentations of the ratios, the line-based presentation is more consistent with a convention—that a ratio is often measured and shown geometrically by the slope of a line rather than the angle of a point or the shape of a geometric space. The line OP, OP', or OP" in the figure is one such line. The slope of each line is equal to the ratio the line represents. It is a geometric measure of the relationship in quantity between the two things compared in the ratio.

The ratio represented by the shaded rectangle, the point P, or the line OP is positive but less than 1 since $D > N$. The ratio represented by the unshaded rectangle, the point P', or the line OP' is also positive but greater than 1 since $D < N$. The ratio represented by the dashed-line square, the point P", or the line OP" is positive and equal to 1 since $D = N$. And the slope of this line is 1, and its angle with the divisor (and dividend) coordinates is 45°. Hereafter, a rectangle representing a ratio such as one in the figure is referred to as the ratio rectangle, and a square representing a ratio like one there as the unit ratio square. And the diagonal (line) of the ratio rectangle is referred to as the ratio line. The diagonal (line) of the unit ratio square in particular is referred to as the unit ratio line. And the ratio represented by the unit ratio square or the unit ratio line is referred to as the unit ratio.

For every ratio including the unit ratio, the location of its ratio rectangle or square and its ratio line in the two-dimensional space is determined by the sign (+ or −) as well as magnitude of D and N. They lie in QI or QIII of the two-dimensional space if both D and N have the same sign (+ or −). Otherwise, they lie in QII or QIV. Regardless, for a given ratio, the three geometric measures—the direction of the point, the slope of the line, and the shape of the space—are equivalent since they all represent the same ratio and the same relationship between the two things compared in the ratio. If one of them is known or determined, therefore, the others are also known and determined.

As ratios are scalar numbers, what really matters to the magnitude of a ratio is the shape (not size) of the ratio rectangle or square representing the ratio, the slope (not length) of the ratio line, and the angle or direction (not distance) of the diagonal vertex (point) of the ratio rectangle or square from the point of origin. Hence, the ratio per se remains the same even if the size of the ratio rectangle or square is changed without changing its shape, the length of the ratio line is changed without changing its slope, and the distance of the vertex from the point of origin is changed without changing its direction.

While different in their geometric presentations, algebraically, all ratios are defined in one simple form as

$$r = N/D$$

where r is the ratio in question. If D, N > 0 and D > N, the ratio r is positive and less than 1 (0 < r < 1). Ratios of this type are represented by a ratio rectangle such as the shaded one in Fig. 9.2. The angle of the vertex (point P) from the point of origin is less than 45°, and the slope of the ratio line OP is less than 1. If D, N > 0 and D < N, the ratio r is positive and greater than 1 (r > 1). Ratios of this type are represented by a ratio rectangle such as the unshaded rectangle in the figure. The angle of the point P' from the point of origin is greater than 45° (but less than 90°), and the slope of the ratio line OP' is greater than 1. If D, N > 0 and D = N, the ratio r is +1. Thus, the ratio line has an angle of 45° with the divisor coordinate and has a slope of 1. Ratios of this type are represented by a unit ratio square such as the dashed-line square in the figure or by the unit ratio line OP".

There can be an almost countless number of ratio rectangles, unit ratio squares, and ratio lines in the two-dimensional space. What this means is that an equally countless number of unique ratios can be presented geometrically in the entire two-dimensional spacespace. This is true especially if both the magnitude of D and N and their sign (+ or −) are considered.

As every ratio can be measured by the slope of its ratio line, it can also be shown by tangent θ—the tangent of the angle of the ratio line with the divisor coordinate. This is possible because the ratio line divides the related ratio rectangle or the square into two contrasting yet symmetric right triangles of the same size. Given this, the ratio can also be measured and shown by tangent θ. It is less than, equal to, or greater than 1, depending on whether 0° < θ < 45°, θ = 45°, or 45° < θ < 90°. Thus, each of the ratios in Fig. 9.2 can be measured and shown as follows:

$$\begin{aligned} r &= N/D = |DP|/|OD|, |D'P'|/|OD'|, \text{or } |D'P''|/|OD'| \\ &= |OP|\text{sine } \theta/|OP|\text{cosine } \theta, |OP'|\text{sine } \theta/|OP'|\text{cosine } \theta, \\ &\quad \text{or } |OP''|\text{sine } \theta/|OP''|\text{cosine } \theta \\ &= \text{sine } \theta/\text{cosine } \theta \\ &= \text{tangent } \theta \end{aligned}$$

where θ is the angle of the ratio line with the divisor coordinate (0° < θ < 90°).

The unit ratio represented by the unit ratio line or unit ratio square has special features. It is +1 or −1, depending on the sign (+ or −) of D and N. And the unit ratio line has an angle of 45° with the divisor *and* dividend coordinates. Thus, the slope of the ratio line is 1 in absolute terms. What this means is that the quantitative relationship between the two things compared in the unit ratio is a perfectly balanced one, although the direction of this relationship can be positive or negative, depending on the sign (+ or −) of D and N. Such a unit ratio is not uncommon in reality. For example, the ratio of the number of fingers between the left and right hands of a person, the ratio of men to women among married couples, and the ratio of the number of players between two sport teams are all unit ratios. Having such features, the unit ratio can be used as a basis for defining various types of ratios.

Types of Ratios

The two rectangles and one square in Fig. 9.2 represent only three different types of ratios that can exist. These ratios have one thing in common—that they are positive as their divisor and dividend both are positive (D, N > 0). Thus, the ratio rectangles, unit ratio square, and ratio lines all lie in QI of the two-dimensional space. There are other types of ratios as well, some of which can be negative. As an extension, Fig. 9.3 geometrically presents all possible types of ratios that can exist and be shown in the divisor-dividend geometric space, including those three types seen in the previous figure. There are two ratio rectangles and one unit ratio square, thus the three ratio lines, in each quadrant of the two-dimensional space. Each of them represents a distinct type of ratio. In total, 12 different types of ratios are presented in the figure.

The 12 different types of ratios can be defined relative to the unit ratio as a benchmark, which is +1 or −1. The ratios that are less than 1 in absolute magnitude ($|r| < 1$) are represented by the ratio rectangle with $|D| > |N|$. There is one such ratio rectangle in each quadrant of the two-dimensional space. The ratio represented by each rectangle is also represented by the angle of the diagonal vertex (point) of the rectangle from the point of origin, the slope of the ratio line, or the shape of the rectangle. The ratios that are greater than 1 in absolute magnitude ($|r| > 1$) are represented by the ratio rectangle with $|D| < |N|$. There is one such ratio rectangle in each quadrant. The ratio represented by each rectangle is also represented by the angle of the diagonal vertex (point) of the rectangle, the slope of the ratio line, or the shape of the ratio rectangle. The ratios that are equal to 1 ($|r| = 1$) are represented by the unit ratio square with $|D| = |N|$. There is one such square in each quadrant. The ratio represented by each square is also represented by the angle of the diagonal vertex (point) of the square or the slope of the unit ratio line.

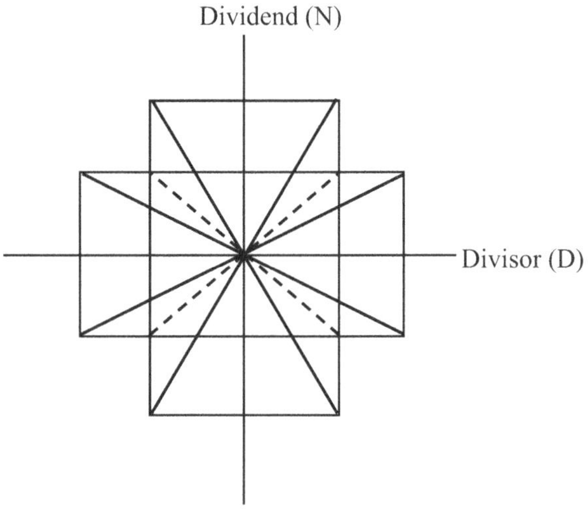

Fig. 9.3 Presenting all possible types of ratios in the two-dimensional geometric space

Shown below are the 12 different types of ratios defined in algebraic form and determined based on the unit ratio as a benchmark.

$$r = +N/+D \text{ or } -N/-D \text{ where } r <, =, > 1$$

or

$$r = -N/+D \text{ or } +N/-D \text{ where } r >, =, < -1$$

The ratio that is equal to zero ($r = 0$) is a trivial one because it means that the dividend of the ratio is zero, which in turn means that there is nothing that is compared with the thing represented by the divisor of the ratio. If the divisor of a ratio is zero, the ratio is not defined mathematically. Again, what this means is that there is nothing that is compared with the dividend of the ratio.

9.2 Financial Ratios and Geometry

The Concept and Nature of Financial Ratios

The discussions of ratios in general in the previous section apply to financial ratios, except that the numbers used to generate them are financial numbers. These financial numbers are generated in accounting. A financial ratio is a number that represents the monetary measure of the quantity of one financial item measured against the monetary measure of the quantity of another financial item. Like ratios in general, financial ratios are scalar numbers and can be positive or negative. And they are relative, in nature.

Financial ratios are generated by a pair of financial numbers, one for D (divisor) and the other for N (dividend). The financial numbers D and N are derived, not primitive, monetary measures in accounting. They are the level 2 or level 3 aggregates of monetary measures discussed in Chap. 8. No financial ratio is generated with the primitive measures—debits and credits—of individual accounts, which are measured and recorded when accounting events involving those accounts take place and are recorded according to the double entry principle.

In order for a financial ratio to be meaningful, the two financial items compared as its dividend and divisor must have a common attribute, and this attribute should be measured and shown in the same monetary unit. If the two financial items have no such common attribute or if they are measured in different monetary units, no meaningful financial ratio can be defined between the two financial items compared. The common attribute is their financial value, and the measurement unit used to measure and show it is the same monetary unit. If, for example, the current ratio of a firm is generated by dividing its current assets by its current liabilities, the financial value of both the current assets and the current liabilities should be measured and shown in the same monetary unit—US dollar, for example, not a mix of both US dollar and euro.

Financial Ratios as Points, Lines, and Spaces

As with ratios in general, financial ratios can be examined geometrically in the two-dimensional (divisor-dividend) Cartesian space with two coordinates, one for the divisor of the ratios and the other for the dividend. The divisor D of a financial ratio is a point or line on the divisor coordinate of the two-dimensional space, and the dividend N is a point or line on the dividend coordinate. And the financial ratio itself is shown by another point or line, or even a space. This space is a rectangle or square as the divisor and dividend coordinates are orthogonal. Unlike the point or line for D and N, the point or line for the financial ratio lies *off*, not on, the divisor and dividend coordinates.

In the two-dimensional Cartesian space, a financial ratio is determined by the angle (direction, not distance) of a point from the point of origin, the slope (not length) of a line, or the shape (not size) of a space (rectangle or square). And each of these is defined by the pair of points or lines representing D and N. The geometric space defined jointly by D and N is a rectangle or square. Hereafter, the former is called the financial ratio rectangle, equivalent to the ratio rectangle in the previous section, and the latter is called the unit financial ratio square, equivalent to the unit ratio square there. Consistently, the diagonal of the rectangle, referred to as the ratio line before, is called the financial ratio line. And the diagonal of the square, referred to as the unit ratio line before, is called the unit financial ratio line. Both the ratio square and the related ratio line represent the same unit financial ratio, which is 1 in absolute value. The space defined by D and N is a rectangle if $|D| \neq |N|$ or a square if $|D| = |N|$. It cannot be any other quadrilateral shape such as parallelogram and trapezoid as D and N are points on the divisor and dividend coordinates that are orthogonal.

When financial ratios are defined geometrically as above, what matters to the magnitude of the ratios is the shape (not size) of their ratio rectangle or square, the slope (not length) of their ratio line, or the direction (or angle) of the diagonal vertex (point) of the rectangle or square from the point of origin. For a given financial ratio, the alternative geometric presentations are equivalent as they all represent the same financial ratio. Hence, if one of them is known or determined, the others are also known or determined. As with ratios in general, the location of the rectangle or square, and the line and point, of financial ratios in the two-dimensional space depends on the signs (+ and −) of D and N. They lie in QI or QIII if D and N have the same sign (+ or −). Otherwise, they lie in QII or QIV.

The unit financial ratio is a special kind of financial ratio as it is 1 in absolute value. Geometrically, it is represented by the unit financial ratio square or the diagonal of this square. What this means is that the relationship between the two financial items compared in the unit financial ratio is a perfectly balanced one, although it can be positive or negative, depending on the sign of D and N. Having such a unique feature, the unit financial ratio can be a basis for determining various types of financial ratios, just as a unit ratio can be such a basis for determining various types of ratios in general as seen before.

Types of Financial Ratios

There are different types of financial ratios. One way to determine them is to use the unit financial ratio as a benchmark. In addition, there are other ways to define types of financial ratios. One is to check where the financial numbers used to generate them come from, and another is to see how financial ratios are actually used for various financial analyses. These two are traditional ways of determining the types of financial ratios.

Types of Financial Ratios Based on the Unit Financial Ratio When the unit financial ratio is considered a benchmark, any of the following can be a type of ratios:

$$r <, =, > +1 \text{ where } r = +N/+D \text{ or } -N/-D$$

and

$$r >, =, < -1 \text{ where } r = +N/-D \text{ or } -N/+D$$

or, in absolute terms,

$$\begin{aligned} |r| &> 1 \rightarrow |N| > |D| \\ &= 1 \rightarrow |N| = |D| \\ &< 1 \rightarrow |N| < |D| \end{aligned}$$

As shown, there are as many as 12 different types of financial ratios, which are equivalent to the 12 different types of ratios in general seen in the previous section. When considered in absolute terms, those 12 types of financial ratios are reduced to three types. By nature, financial ratios cannot be zero ($|r| \neq 0$) for the reason indicated earlier with respect to ratios in general. The most common types of financial ratios are those that are defined with +D and + N, which are the level 2 or level 3 aggregates of monetary measures in accounting (see Sect. 8.2). These aggregates are positive unless the firm (business entity) in question has financial or other difficulties. Such financial ratios are positive and can be greater than, equal to, or less than +1, depending on whether +D <, =, or > +N.

What the above means geometrically is that, for most firms, the financial ratio rectangle, the unit financial ratio square, and the related financial ratio line all lie in QI of the two-dimensional divisor-dividend space. The exact shape of the ratio rectangle depends on its length versus width, that is, +D versus +N. If +D > +N, the length of the ratio rectangle is longer than its width and the financial ratio is positive and less than 1 ($0 < r < +1$). The slope of the financial ratio line is less than 1, and the angle of the ratio line with the divisor coordinate is less than 45° ($\theta < 45°$). Examples of such financial ratios are the net income-to-total equity ratio (rate of return on book equity) and the gross margin (gross margin-to-total sales) ratio. Conversely, if +D < +N, the length of the ratio rectangle is shorter than its width and the financial ratio is positive and greater than one ($r > +1$). The slope of the financial ratio line is larger than 1, and the angle of this line with the divisor coordinate is greater than 45° but less than 90° (45° < θ < 90°). Examples of such financial ratios include the inventory asset turnover (total sales-to-inventory assets) ratio and the interest coverage (operating income-to-interest expense) ratio.

Although not common, financial ratios for some firms can be negative (r < 0) as their divisor or dividend is negative (−D or −N). In this case, they can be greater or less than, or even equal to, −1. Such financial ratios are not common, again because D and N are the level 2 or level 3 aggregates of monetary measures, which are positive for most or normal firms. Yet such negative financial ratios can exist for firms having financial or other difficulties, or for some accounting periods even for normal firms. If D > 0 and N < 0, the financial ratio is negative and can be greater or less than −1, depending on whether |D| < or > |N|. In this case, the financial ratio rectangle and the financial ratio line both lie in QIV of the two-dimensional space. Such financial ratios typically involve net loss, negative working capital, and negative cash flows, among others. Examples include the net loss-to-total assets ratio, the net loss-to-book equity ratio, and the net loss-to-sales ratio. Conversely, if D < 0 and N > 0, the financial ratio is also negative and can be greater or less than −1, depending on whether |D| < or > |N|. For such financial ratios, both the ratio rectangle and the ratio line lie in QII of the two-dimensional space. Financial ratios of this type typically involve the negative book value of equity. Examples include the negative debt-to-equity ratio and the negative rate of return on book equity, for firms that have long-term debt and net income but a negative book value of owner equity.

Although rare, some financial ratios of a firm may be defined with both −D and −N if the firm has severe financial or other difficulties.[1] If so, the financial ratios are positive even though their divisor and dividend are negative. For example, a firm may have severe financial or other difficulties so that it has a negative owner equity on books as well as net loss for an accounting period. If so, the book rate of return of the firm is positive as it is defined by dividing −N by −D. The same is also true if a firm has negative free cash flows as well as a negative owner equity. In this case, the free cash flow-to-book equity ratio of the firm is also positive. For such financial ratios, the ratio rectangle, square, and related ratio line all lie in QIII of the two-dimensional space. They are rare, again because it is rare that a firm has both D and N as negative level 2 or level 3 aggregates of monetary measures.

There is still another type of financial ratios that is not common. This is unit financial ratios. Although most financial ratios are not of this type, by nature, some must be of the unit type, even for normal firms. Examples are the ratio of total assets to total liabilities plus owner equity for every firm and the ratio of total owner equity

[1]If such difficulties persist over a long period of time, the firm is unlikely to continue to exist as a "going-concern." Conceptually, such firms are not considered normal. Nor is their existence consistent with the going-concern assumption in accounting.

to total assets for a firm with no liability as part of the sources of its capital resources. Conversely, some other financial ratios cannot be of the unit type, by nature. Examples are the net margin (net profits-to-total revenue) ratio and the net loss-to-total assets ratio. While a financial ratio can be of the unit type and positive, a negative unit financial ratio defined with both −D and −N is rare. Again, this is because D and N are a level 2 or level 3 aggregate of individual accounts or a group of accounts and such aggregates are (supposed to be) positive for most firms.

Types of Financial Ratios Based on the Sources of Financial Numbers Besides using the unit financial ratio as a benchmark, there is another way to determine the types of financial ratios—to check where the financial numbers used to generate them come from. When the sources of those numbers are used as a basis for determining the types of financial ratios, there are four different types of financial ratios:

(a) The balance sheet-based ratios (e.g., the current ratio and the debt-to-equity ratio)
(b) The income statement-based ratios (e.g., the gross margin ratio and the net income-to-total revenue ratio)
(c) The cash flow statement-based ratios (e.g., the cash flows from operation-to-total cash flows ratio and the free cash flows to net cash flows ratio)
(d) The inter-statement ratios (e.g., the inventory turnover ratio, the return-on-equity ratio, and the return-on-total assets ratio).

The balance sheet-based financial ratios are generated with financial numbers from balance sheet accounts only, such as the current ratio and the debt-to-equity ratio. The income statement-based financial ratios are generated with financial numbers from income statement accounts only, such as the net margin or gross margin ratio. And the cash flow statement-based financial ratios are generated with financial numbers from the cash flow statement, such as the net working capital-to-total cash flows ratio. The inter-statement financial ratios are defined with financial numbers from different financial statements, such as the inventory asset turnover ratio and the return on equity ratio.

Types of Financial Ratios Based on Their Usage There is still another way of determining the types of financial ratios. It is based on how financial ratios are actually used for financial analysis. When determined this way, financial ratios belong to one of the following types:

(a) The liquidity ratios (e.g., the current assets-to-current liabilities ratio and the quick assets-to-current liabilities ratio)
(b) The profitability ratios (e.g., the net income-to-total revenues ratio and the net income-to-book equity ratio)
(c) The activity (or efficiency) ratios (e.g., the inventory asset turnover ratio and the total assets turnover ratio)

(d) The leverage ratios (e.g., the debt-to-equity ratio and the debt-to-total assets ratio)
(e) Others.

9.3 Ratios and Financial Ratios in Vector Spaces

Ratios and Ratio Vectors in General

When presented geometrically, a ratio is shown by a point or line or even a space, and each of these is determined by D and N for the ratio. Since a point, line, and space respectively are a vector or a set of vectors in a vector space, the point, line, or space representing a ratio can also be a vector or a set of vectors. A vector is an array of real numbers. So is a vector for a ratio. This vector, referred to as the ratio vector hereafter, is defined with a pair of real numbers, D and N. It is a two-element vector or a vector of order 2. It can be a row or column vector, depending on how the pair of numbers is arranged, row- or column-wise. The vector space where ratio vectors lie is a two-dimensional (divisor-dividend) space with two coordinates. It has an origin—the point at which the divisor and dividend coordinates intersect. In this space, the vectors representing ratios are all position vectors and start at the origin.

For every ratio, D and N each are a point or line, hence a vector, on the divisor and dividend coordinates respectively of the two-dimensional vector space. They are coordinate vectors of order 2, denoted as **D** and **N** hereafter. **D** is called the divisor vector and **N** the dividend vector. They are written as

$$\mathbf{D} = (+\mathrm{D}, 0) \text{ or } (-\mathrm{D}, 0)$$

$$\mathbf{N} = (0, +\mathrm{N}) \text{ or } (0, -\mathrm{N})$$

As coordinate vectors, **D** and **N** are mutually orthogonal and conformable for vector addition.

As a pair, **D** and **N** can be basis vectors and generate another vector. This vector, denoted as **r**, is referred to as the ratio vector hereafter. It is a sum vector of order 2, generated as a linear combination of **D** and **N** as shown below.

$$\begin{aligned}\mathbf{r} &= \mathbf{D} + \mathbf{N} \\ &= (+\mathrm{D}, +\mathrm{N}), (+\mathrm{D}, -\mathrm{N}), (-\mathrm{D}, +\mathrm{N}), \text{or } (-\mathrm{D}, -\mathrm{N})\end{aligned}$$

The ratio vector **r** can be one of the four types shown, depending on the sign of D and N. As a vector, **r** is not the same as the ratio itself ($\mathbf{r} \neq \mathrm{r}$). But it represents the ratio r. It is equivalent to the ratio line in Sect. 9.1. And the ratio r is equal to the slope of **r**, just as a ratio in general is equal to the slope of the ratio line, not the ratio line itself. **r** is equal to the diagonal of the ratio rectangle or square, which is defined by **D** and **N**. It is a diagonal vector.

Given **D** and **N**, the ratio is defined as

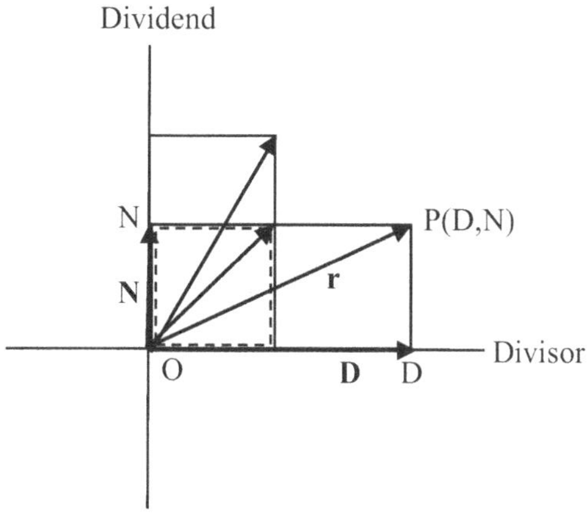

Fig. 9.4 Presenting different types of ratios in the two-dimensional vector space

$$\begin{aligned} r &= |\mathbf{N}|/|\mathbf{D}| \\ &= +N/+D,\ +N/-D,\ -N/+D, \text{or} -N/-D \end{aligned}$$

As **r** is not the same as the ratio r, nor is its magnitude (length) equal to the ratio ($|\mathbf{r}| \neq |r|$). Rather, the ratio is equal to the slope (or angle) of **r**, which is measured as shown above. If, for example, $\mathbf{D} = (3, 0)$ and $\mathbf{N} = (4, 0)$, the ratio r is equal to roughly 1.33 as $|\mathbf{D}| = 3$ and $|\mathbf{N}| = 4$. Since vectors have a direction, dividing one vector by another vector does not give a ratio.[2] However, the magnitude of a vector can be divided by the magnitude of another vector, which is what is shown above. Given this, the following holds:

$$r <, =, \text{or} > 1 \text{ if } |\mathbf{D}| >, =, < |\mathbf{N}|$$

Figure 9.4, adapted from Fig. 9.2, shows three different types of ratio vectors in the divisor-dividend vector space. They are shown by the diagonal arrows, and each of them is a distinct ratio vector. There are two rectangles and one square in the figure. Thus, there are three different types of ratio vectors. For each ratio vector, D and N are assumed to be positive (D, N > 0) so that **D** and **N** lie on the positive side of the respective coordinates. As a pair, **D** and **N** can be basis vectors, define the ratio vector **r**, and span the shaded ratio rectangle. There is a similar ratio vector relating to the unshaded ratio rectangle and the dashed-line unit ratio square respectively (although it is not shown to keep the figure simple). The shaded ratio rectangle represents a ratio that is less than 1 as +D > +N and $|\mathbf{D}| > |\mathbf{N}|$. And **r** represents the ratio. The slope of **r**, not its magnitude $|\mathbf{r}|$, is equal to the ratio r. The angle of **r** with the divisor coordinate is less than 45° ($\theta < 45°$) so that the slope of **r**, namely the ratio r, is positive and less than 1 ($0 < r < +1$).

[2]Since vectors have a direction, dividing one vector by another is like dividing one direction by another (e.g., dividing the left side by the right side). The result has no meaningful interpretation, even conceptually.

While their orientation is different, the unshaded ratio rectangle and the related ratio vector are interpreted similarly. Both of them represent the same ratio, which has +D < +N and |**D**| < |**N**|. The ratio vector **r** has an angle greater than 45° (45° < θ < 90°) with the divisor coordinate, and the slope of **r**, namely the ratio r, is positive and greater than 1 (r > +1). The dashed-line square represents the unit ratio that has +D = +N and |**D**| = |**N**|. And **r** for this ratio has an angle of 45° with the divisor coordinate (θ = 45°) so that it has a slope of 1. This ratio vector, referred to as the unit ratio vector hereafter, is equivalent to the unit ratio line in Sect. 9.1.[3]

As coordinate vectors, **D** and **N** for each type of the ratios above can be basis vectors. As a pair, they generate not only the ratio vector **r** but also span a vector space. This space is a rectangle or square, depending on whether |**D**| >, =, or < |**N**|. It cannot be any other type of polygon such as parallelogram or rhombus since **D** and **N** are orthogonal.

The ratio vector **r** divides the related ratio rectangle or square into two contrasting yet symmetric right triangles of equal size as shown in the figure. Given this, the ratio r can also be shown by tangent θ:

$$\begin{aligned} \mathrm{r} &= |\mathbf{N}|/|\mathbf{D}| = |\mathbf{r}|\text{sine } \theta/|\mathbf{r}|\text{cosine } \theta \quad (0^{\circ} < \theta < 90^{\circ}) \\ &= \text{tangent } \theta \end{aligned}$$

where θ is the angle between **r** and **D**. The ratio r is equal to the slope of **r**, not its magnitude (r ≠ |**r**|), which is measured by tangent θ. If, for example, **D** = (3, 0) and **N** = (4, 0), θ ≈ 53.13° and tangent θ = 1.33. Hence, r = 1.33. The ratio is less than 1 (|r| < 1) if 0° < θ < 45° (|**D**| > |**N**|) and greater than 1 (|r| > 1) if 45° < θ < 90° (|**D**| < |**N**|). And it is equal to 1 (r = 1) if θ = 45° (|**D**| = |**N**|), in which case tangent θ = 1 as sine θ = cosine θ.

Figure 9.5 extends what Fig. 9.4 shows. It shows all possible types of ratio vectors that can exist in the two-dimensional vector space. The diagonal arrow in each quadrant of the vector space is a ratio vector. The related coordinate vectors **D** and **N** are not shown to keep the figure simple. As seen, there are three different types of ratio vectors in each quadrant and, thus, a total of 12 different types of ratio vectors across the four quadrants. They are equivalent to the 12 different types of ratios in general seen in Fig. 9.3 before.

Regardless of the types of ratio vectors, what matters to the ratios is the slope, not the length or magnitude, of **r**. That is, what determines a ratio is the slope or direction (not distance or length) of **r** or the shape (not size) of the ratio rectangle or square. The slope of **r** depends on the magnitude and direction of **D** and **N** and their interior angle. For a given ratio, the alternative geometric presentations of the ratio are equivalent. Thus, if one of them is known or determined, the others are also known or determined.

[3]The unit ratio vector here does not mean that **r** = (1, 1). Nor is it a unit vector in vector theory, which is defined as a vector whose magnitude is 1. A vector whose elements are all 1 is not a unit vector. See Sect. 6.7, Chap. 6.

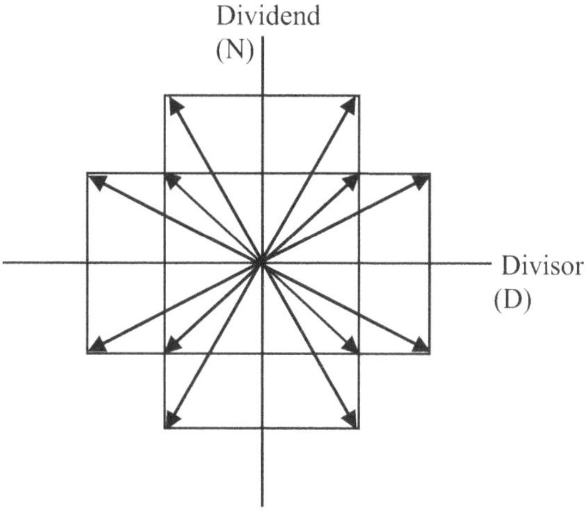

Fig. 9.5 Presenting all possible types of ratios in the two-dimensional vector space

Financial Ratios in Vector Spaces

The discussions thus far of ratios in vector form and their presentations in the divisor-dividend vector space apply to financial ratios. The vectors **D** and **N** for ratios before are now the divisor and dividend coordinate vectors for financial ratios and represent financial numbers. The ratio vector **r** before is the financial ratio vector. And the ratio rectangle before is the financial ratio rectangle, and the unit ratio square is the unit financial ratio square. **D** and **N** can be a pair of basis vectors, generate **r**, and span the financial ratio rectangle or square. The financial ratio vector **r** is a linear combination of **D** and **N**, and the financial ratio r is equal to the slope of **r**. These are shown below.

$$\begin{aligned} \mathbf{r} &= \mathbf{D} + \mathbf{N} \\ \mathrm{r} &= |\mathbf{N}|/|\mathbf{D}| \\ &= \text{tangent } \theta \end{aligned}$$

where θ is the angle between **r** and **D** ($0° < \theta < 90°$). As with ratios in general, the financial ratio r is not defined as **N / D**, nor equal to the magnitude of **r**, again because vectors have a direction so that dividing **N** by **D** does not give a ratio. Yet the magnitude of a vector can be divided by the magnitude of another vector, which is what is shown above. **r** is a diagonal vector, equal to the diagonal of the financial ratio rectangle or square defined by **D** and **N**.

As with ratio vectors in general, the financial ratio vector **r** itself is not the same as the financial ratio r. Nor is its magnitude the same as r ($|\mathbf{r}| \neq \mathrm{r}$). Instead, the slope of **r** is the same as r, which is measured by tangent θ. If $\theta = 45°$ in particular, tangent $\theta = 1$, meaning that **D** and **N** are orthogonal and $|\mathbf{D}| = |\mathbf{N}|$. And **r** is a unit financial ratio vector, the slope of **r** is 1, and the ratio is a unit financial ratio. It means that the relationship between the two financial items compared in the financial ratio is a perfectly balanced one although this relationship can be positive or negative, depending on the direction of **D** and **N**—the sign (+ or −) of D and N. Having such unique features, the unit financial ratio vector can be a basis for determining various types of financial ratio vectors, just as the unit ratio can be a basis for determining various types of ratios in general (see Sect. 9.1).

Given the unit financial ratio as a benchmark, every financial ratio belongs to one of the following three types (in absolute terms):

$$\begin{aligned} r &= |\mathbf{N}|/|\mathbf{D}| = \text{tangent}\,\theta < 1 \\ &= |\mathbf{N}|/|\mathbf{D}| = \text{tangent}\,\theta = 1 \\ &= |\mathbf{N}|/|\mathbf{D}| = \text{tangent}\,\theta > 1 \end{aligned}$$

Again, no financial ratio can be of the type: $|\mathbf{N}| / |\mathbf{D}| = 0$ as the financial numbers D and N are level 2 or level 3 aggregates of monetary measures, which are (supposed to be) positive as indicated earlier. If $|\mathbf{N}| = 0$ and thus $r = 0$, this means that there is no financial item that is compared with the financial item represented by D.

The three types of financial ratios above are defined based on the magnitude of **N** and **D**. If the direction (+ or −) of **D** and **N** is considered as well, there can be as many as 12 different types of financial ratio vectors, which are equivalent to the 12 different types of ratio vectors in general seen earlier. The different types of ratio vectors in Fig. 9.5 now become 12 different types of financial ratio vectors in the two-dimensional vector space. The location of both **r** and the related ratio rectangle or square for each type of financial ratios depends on the sign (+ or −) of D and N. If D and N have the same sign (+ or −), **r** and the related ratio rectangle or square lie in QI or QIII of the two-dimensional vector space. Otherwise, they lie in QII or QIV.

For most financial ratios, D and N are positive numbers since they are level 2 or level 3 aggregates of monetary measures—the period-end balance (debit or credit) of individual accounts or a group of accounts as indicated before. Therefore, **r** and the related ratio rectangle or square lie in QI. If +D > +N, the slope of **r** is less than 1, meaning that the financial ratio is positive and less than 1 ($0 < r < 1$). Examples of such financial ratios are the gross margin ratio, the return on total assets ratio, the return on equity ratio, and the current-to-total assets ratio. If +D < +N, the slope of **r** is greater than 1, meaning that the financial ratio is positive and greater than 1 ($r > 1$). Examples of such financial ratios are the accounts receivable turnover ratio, the inventory asset turnover ratio, the total assets to equity ratio, and the operating assets-to-operating liabilities ratio.

Although not common, some financial ratios can be negative ($r < 0$) since D and N do not have the same sign (+ or −). For such financial ratios, the ratio vector **r** and the related ratio rectangle or square lie in QII or QIV. And the slope of **r** can be greater or less than −1, depending on whether $|D| < |N|$ or $|D| > |N|$. This can be the case if either D or N as a level 2 or level 3 aggregate of monetary measures is a negative number. And such negative financial ratios can exist for a firm that has financial or other difficulties. Examples are the net loss-to-total assets ratio (i.e., negative return on assets or just negative ROA) and the net loss-to-equity ratio (i.e., negative return on owner equity or just negative ROE).

Some financial ratios are of the unit type so that the slope of **r** is equal to 1 and the financial ratio represented by **r** is 1 ($r = 1$). A unit financial ratio vector has interesting features. It has a slope of +1 or −1 as it has an angle of 45° with both **D** and **N**. The location of **r**, namely the location of the unit financial ratio square, in the two-dimensional vector space depends on the sign of D and N; it lies in QI or QIII if both D and N have the same sign (+ or −) or in QII or QIV otherwise. Example of unit financial ratios were given before—the ratio of total assets to total liabilities plus owner equity and the ratio of total assets to total owner equity for firms with no liabilities.

Products of the Financial Ratio-Related Vectors

For every financial ratio, there are three distinct vectors **D**, **N**, and **r**. As with vectors in general, they can be multiplied with each other since they are of order 2 and conformable for vector multiplication. Thus, it is possible to have various forms of their dot products.[4] Given below are three such dot products.

$$\mathbf{D}.\mathbf{N}^{\mathrm{T}} = |\mathbf{D}||\mathbf{N}|\text{cosine } \theta = \mathbf{N}.\mathbf{D}^{\mathrm{T}} = |\mathbf{N}||\mathbf{D}|\text{cosine } \theta = 0$$
$$\text{cosine } \theta = \mathbf{D}.\mathbf{N}^{\mathrm{T}}/|\mathbf{D}||\mathbf{N}| \text{ or } \mathbf{N}.\mathbf{D}^{\mathrm{T}}/|\mathbf{N}||\mathbf{D}| = 0$$

$$\mathbf{D}.\mathbf{r}^{\mathrm{T}} = |\mathbf{D}||\mathbf{r}|\text{cosine } \theta \neq 0$$
$$\text{cosine } \theta = \mathbf{D}.\mathbf{r}^{\mathrm{T}}/|\mathbf{D}||\mathbf{r}| \neq 0$$

$$\mathbf{N}.\mathbf{r}^{\mathrm{T}} = |\mathbf{N}||\mathbf{r}|\text{cosine } \theta \neq 0$$
$$\text{cosine } \theta = \mathbf{N}.\mathbf{r}^{\mathrm{T}}/|\mathbf{N}||\mathbf{r}| \neq 0$$

where θ is the angle between the two vectors being multiplied and the superscript T means a transpose of the (row) vector shown. That is, $\mathbf{N}^{\mathrm{T}}$, $\mathbf{D}^{\mathrm{T}}$ and $\mathbf{r}^{\mathrm{T}}$ are all column vectors. Since the dot product of vectors is commutative, $\mathbf{D}.\mathbf{N}^{\mathrm{T}} = \mathbf{N}.\mathbf{D}^{\mathrm{T}}$ for example. Each dot product above yields a scalar number. The dot product $\mathbf{D}.\mathbf{N}^{\mathrm{T}}$ is zero since the two vectors being multiplied are orthogonal ($\theta = 90°$) and cosine 90° = 0. The same is also true for $\mathbf{N}.\mathbf{D}^{\mathrm{T}}$. However, neither $\mathbf{D}.\mathbf{r}^{\mathrm{T}}$ nor $\mathbf{N}.\mathbf{r}^{\mathrm{T}}$ is zero since the two vectors are not orthogonal ($\theta \neq 90°$).

As with the dot product of vectors in general, the dot products above have geometric interpretations. It indicates how much the two vectors point in similar directions in the two-dimensional (divisor-dividend) vector space. And their directional similarity is measured by cosine θ. Specifically, the dot product $\mathbf{D}.\mathbf{N}^{\mathrm{T}}$ indicates how much **D** and **N** point in similar directions, and this directional similarity is measured by cosine θ. As the angle between **D** and **N** is 90°, cosine 90° = 0. Thus, $\mathbf{D}.\mathbf{N}^{\mathrm{T}} = \mathbf{N}.\mathbf{D}^{\mathrm{T}} = 0$, meaning that **D** and **N** are orthogonal and have no directional similarity. Also, the vector multiplication projects **N** onto **D** (or vice versa), which

[4]For discussions about the dot product of vectors, see Sect. 6.7, Chap. 6.

generates a projection vector ($\mathbf{N}_p$). This vector is a null vector ($\mathbf{N}_p = \mathbf{0}$) as **N** is orthogonal to **D**. It represents the point of origin. As a pair of basis vectors, **D** and **N** span a vector space. This space is a financial ratio rectangle if $|\mathbf{D}| \neq |\mathbf{N}|$ or a unit financial ratio square if $|\mathbf{D}| = |\mathbf{N}|$.

As for the dot product $\mathbf{D}.\mathbf{r}^T$, the geometric interpretations are also similar. It indicates how much **D** and **r** point in similar directions, and this directional similarity is measured by cosine θ. Since the angle between the two vectors is acute ($0° < \theta < 90°$), cosine $\theta \neq 0$ or 1, meaning that the two vectors are not collinear, nor orthogonal, but point in similar directions. Also, the vector multiplication projects **r** onto **D** (or vice versa) and yields a projection vector ($\mathbf{r}_p$). This vector is not a null vector but equal to **D** ($\mathbf{r}_p = \mathbf{D}$) since it is a perpendicular projection of **r** onto **D**. Having an acute interior angle, **D** and **r** span a vector space. This space is a parallelogram, not a rectangle or square, whose base is equal to $\mathbf{r}_P$ (or **D**) and whose height is equal to the difference $|\mathbf{r} - \mathbf{r}_p|$. This difference may be seen as the magnitude of the difference vector $\mathbf{r} - \mathbf{r}_p$. The size of the parallelogram is equal to: $|\mathbf{r}_P| \times |\mathbf{r} - \mathbf{r}_p|$. And it depends on the magnitude of the two vectors and their interior angle. It is equal in size to the financial ratio rectangle or square spanned by **D** and **N**. Given $\mathbf{r}_p$, the following hold:

$$\begin{aligned} \mathbf{r}_P &= \mathbf{r}\,\text{cosine}\,\theta = \mathbf{D} \\ |\mathbf{r}_P| &= |\mathbf{r}||\text{cosine}\,\theta| = |\mathbf{D}| \end{aligned}$$

and

$$\mathrm{r} = |\mathbf{N}|/|\mathbf{r}_P|$$

Since $\mathbf{r}_p = \mathbf{D}$, the financial ratio r can be defined with $|\mathbf{r}_P|$ as shown. The magnitude of $\mathbf{r}_P$ varies with the slope of **r**, namely the angle of **r** with **D** ($0° < \theta < 90°$); the larger the angle, the steeper the slope of **r**, and the smaller is $|\mathbf{r}_P|$. Hence, the larger is the financial ratio r. The financial ratio is inversely related to $|\mathbf{r}_P|$ and positively related to the angle of **r** with **D**. The parallelogram spanned by **D** and **r** is equal in size to the financial ratio rectangle or square spanned by **D** and **N**.

As for the dot product $\mathbf{N}.\mathbf{r}^T$, the vector multiplication is also performed and interpreted similarly. The dot product indicates the directional similarity between **N** and **r**. The vector multiplication projects **r** onto **N** and generates a projection vector ($\mathbf{r}_P$) on the dividend coordinate. This projection vector is identical to **N** ($\mathbf{r}_p = \mathbf{N}$) and its magnitude varies with the slope of **r**, namely the angle of **r** with **N**. And $\mathrm{r} = |\mathbf{r}_P| / |\mathbf{D}|$. The larger the angle, the gentler the slope of **r** (measured relative to the divisor coordinate). And the smaller is $|\mathbf{r}_P|$ and, hence, the smaller is the ratio r. The financial ratio is positively related to $|\mathbf{r}_P|$ and negatively related to the angle of **r** with **N**. Since the angle of **r** with **N** is acute ($0° < \theta < 90°$), the vector space spanned by them is a parallelogram. The orientation of this parallelogram is symmetric to the orientation of the parallelogram spanned by **D** and **r** before. The parallelogram is symmetric in shape to the parallelogram spanned by **D** and **r**. Also, it is equal in size to the financial ratio rectangle or square spanned by **D** and **N**.

9.4 Ratios and Financial Ratios in Matrix Form

Ratios in Matrix Form and the Financial Ratio Matrix

A matrix is a rectangular or square array of real numbers. A matrix representing a ratio is also such an array of real numbers. It is a 2 × 2 diagonal or scalar matrix and has a rank of 2. Hereafter, it is called the ratio matrix and denoted as **R**. D and N are the main diagonal elements of **R**. The geometric setting where ratios can be examined in matrix form is a two-dimensional (divisor-dividend) vector space, similar to the two-dimensional vector space in the previous section. The first column of **R** represents the point D on the divisor coordinate of the vector space, and the second column represents the point N on the dividend coordinate. Each of these points is considered a two-element (column) vector: **D** = (D, 0) and **N** = (0, N). And **R** is a 2 × 2 diagonal matrix, written as

$$\mathbf{R} = [\mathbf{D}\ \ \mathbf{N}] = \begin{bmatrix} \mathrm{D} & 0 \\ 0 & \mathrm{N} \end{bmatrix}$$

If D and N are the same scalar number, positive or negative, **R** is a scalar matrix and is called a unit ratio matrix hereafter.[5] If |D| = |N| in particular, **R** represents a unit ratio, and the ratio is +1 or −1. As a pair of basis vectors, **D** and **N** generate other vectors such as the ratio vector **r** and span a vector space such as the ratio rectangle or the unit ratio square. This vector space is represented by **R**. As a 2 × 2 diagonal matrix, **R** has a (nonzero) determinant, which is equal in absolute value to the size of the ratio rectangle or unit ratio square represented by **R**.

The discussions above of ratios in matrix form in general also apply to financial ratios. **R** is now the matrix of financial ratios, thus referred to as the financial ratio matrix hereafter. It is a 2 × 2 diagonal matrix and has a rank of 2. And its main diagonal elements D and N are financial numbers. If D and N are the same scalar number, positive or negative, **R** is a scalar matrix and the financial ratio represented by **R** is +1. And **R** is the unit financial ratio matrix. Regardless, **R** is always a symmetric matrix since diagonal matrices are part of the symmetric matrix family.

For every financial ratio, the main diagonal elements D and N are the level 2 or level 3 aggregates of the monetary measures of individual accounts or a group of accounts. Thus, they are positive numbers for most firms. Yet they may be negative numbers for some firms, especially those that have financial or other difficulties, or for some accounting periods for some firms for other reasons. Given this possibility, **R** can be written in one of the following four possible forms:

[5] The unit ratio matrix in this chapter is not the same as a unit (identity) matrix in matrix algebra, which is a diagonal matrix whose diagonal elements are all 1. For the definition of a unit matrix, see Sect. 7.1, Chap. 7.

$$\mathbf{R} = [\mathbf{D} \quad \mathbf{N}] = \begin{bmatrix} +\mathrm{D} & 0 \\ 0 & +\mathrm{N} \end{bmatrix} \begin{bmatrix} +\mathrm{D} & 0 \\ 0 & -\mathrm{N} \end{bmatrix} \begin{bmatrix} -\mathrm{D} & 0 \\ 0 & +\mathrm{N} \end{bmatrix} \text{or} \begin{bmatrix} -\mathrm{D} & 0 \\ 0 & -\mathrm{N} \end{bmatrix}$$

The first and last matrices above are a scalar matrix if D and N are the same scalar number, positive or negative. And the financial ratio represented by **R** is the unit financial ratio. The second and third matrices are diagonal, not scalar, matrices as D and N are not the same scalar number.

When considering the sign (+ or −) as well as magnitude of the main diagonal elements of **R**, the various forms of **R** above can be refined further as follows:

$$+\mathrm{N} > +\mathrm{D},\ +\mathrm{N} < +\mathrm{D}, \text{or} +\mathrm{N} = +\mathrm{D}$$

$$-\mathrm{N} > -\mathrm{D},\ -\mathrm{N} < -\mathrm{D}, \text{or} -\mathrm{N} = -\mathrm{D}$$

$$|+\mathrm{N}| > |-\mathrm{D}|, |+\mathrm{N}| < |-\mathrm{D}|, \text{or } |+\mathrm{N}| = |-\mathrm{D}|$$

$$|-\mathrm{N}| > |+\mathrm{D}|, |-\mathrm{N}| < |+\mathrm{D}|, \text{or } |-\mathrm{N}| = |+\mathrm{D}|$$

Each pair of the main diagonal elements above, whether in absolute value or not, represents a form of **R** and, hence, a distinct financial ratio. For example, the first pair +N > +D above represents **R** with +N and +D as its diagonal elements. And **R** represents a financial ratio that is greater than +1. Similarly, the last pair |−N| = |+D| above means **R** with −N and +D as its diagonal elements. And **R** represents a financial ratio that is equal to −1. When defined this way, there can be as many as 12 different forms of **R**, which are equivalent to the 12 different types of financial ratios or financial ratio vectors seen earlier.

Geometrically, a matrix represents a vector space, which is spanned by the columns (or rows) of the matrix. This is also true for **R**. The vector space represented by **R** is spanned by **D** and **N**. These vectors are mutually orthogonal and, as a pair of basis vectors, span the vector space. This vector space is a financial ratio rectangle if |**D**| ≠ |**N**| or a unit financial ratio square if |**D**| = |**N**|. The diagonal of this vector space is equal to the financial ratio vector **r**, and the slope of **r** is equal to the financial ratio r. If **R** is a scalar matrix in particular, it is a unit financial ratio matrix and the vector space represented by **R** is a unit financial ratio square. And the ratio is a unit financial ratio. This is so, regardless of whether the diagonal elements D and N are positive (+) or negative (−), as long as they both have the same sign and the same magnitude. In this case, **R** can be written as **R** = k**I,** where k is the scalar number and **I** is a 2 × 2 identity matrix. The four vertices of the unit financial ratio square represented by **R** can be determined by multiplying D (or N) as the scalar number k with each of the column vectors: (0, 0), (1, 0), (0, 1), and (1, 1). Together, these column vectors define a square whose size is 1.

In general, the location of the financial ratio vector **r**, hence the location of the financial ratio rectangle or square, in the two-dimensional vector space depends on the sign (+ or −) of the main diagonal elements of **R** as shown below.

$$
\begin{array}{ll}
+\mathrm{N} > +\mathrm{D} \text{ or } +\mathrm{N} < +\mathrm{D} & \rightarrow \mathbf{r} \text{ in QI} \\
+\mathrm{N} > -\mathrm{D} \text{ or } +\mathrm{N} < -\mathrm{D} & \rightarrow \mathbf{r} \text{ in QII} \\
-\mathrm{N} > -\mathrm{D} \text{ or } -\mathrm{N} < -\mathrm{D} & \rightarrow \mathbf{r} \text{ in QIII} \\
-\mathrm{N} > +\mathrm{D} \text{ or } -\mathrm{N} < +\mathrm{D} & \rightarrow \mathbf{r} \text{ in QIV} \\
+\mathrm{N} = +\mathrm{D} \text{ or } -\mathrm{N} = -\mathrm{D} & \rightarrow \mathbf{r} \text{ in QI or QIII} \\
|+\mathrm{N}| = |-\mathrm{D}| \text{ or } |-\mathrm{N}| = |+\mathrm{D}| & \rightarrow \mathbf{r} \text{ in QII or QIV}
\end{array}
$$

Features of the Financial Ratio Matrix

The financial ratio matrix **R** has interesting features. It is a 2 × 2 diagonal or scalar matrix. Thus, it is always a 2 × 2 symmetric matrix, meaning $\mathbf{R} = \mathbf{R}^{\mathrm{T}}$. Interchanging its rows and columns does not change its features. Nor does this affect its determinant as shown below.

$$
|\mathbf{R}| = \begin{vmatrix} \mathrm{D} & 0 \\ 0 & \mathrm{N} \end{vmatrix} = |\mathrm{D.N} - 0.0| = |\mathrm{D.N}| = \left|\mathbf{R}^{\mathrm{T}}\right| > 0
$$

The determinant |**R**| in absolute terms is equal to the size of the financial ratio rectangle or square represented by **R**. And interchanging the rows and columns of **R** does not affect the size of this rectangle or square as **R** is a symmetric matrix.

Besides, **R** has other interesting features when it is compared to the double entry (DE) matrix **M** for accounting events seen in Chap. 7. They both are a diagonal or scalar matrix and have the features and properties of diagonal or scalar matrices in general. In this regard, they are similar. However, **R** differs from **M** in other respects. First of all, it concerns financial ratios, whereas **M** concerns accounting events. As such, it is defined in a different geometric setting, namely in the divisor-dividend vector space, whereas **M** is defined in the debit-credit vector space. Hence, **R** has nothing to do with the double entry principle or the notion of debit and credit. Nor does it have anything to do with the duality and balance features of accounting events or double entry accounting in general. And the diagonal elements of **R** are the level 2 or level 3 aggregates (derived measures) of the monetary measures of individual accounts or groups of similar accounts. But the main diagonal elements of **M** are not such aggregates; they are the primitive monetary measures (debits and credits) of individual accounts. They are generated and recorded as the accounting events involving those accounts take place and are recorded according to the double entry principle. As such, they are yet to be aggregated. The vector space represented by **R** is a rectangle or square in the two-dimensional divisor-dividend vector space. For most financial ratios, the vector space is a rectangle rather than a square since it is not common that D = N or |D| = |N|. However, the vector space represented by **M** is always a square—a DE square—in the two-dimensional debit-credit vector space. This is because every accounting event has duality and balance as its fundamental features so that its debit entry should match its credit entry. And **M** reflects these features or, more generally, the features of double entry accounting.

Although different from **M** in those respects above, **R** is similar in some other features to **M** when it is seen mathematically. For example, the transpose of **R** as a diagonal matrix is the same as **R** itself ($\mathbf{R}^T = \mathbf{R}$), which is also true for **M** ($\mathbf{M}^T = \mathbf{M}$). **R** has a nonzero determinant ($|\mathbf{R}| \neq 0$). So does **M** (see Sect. 7.2, Chap. 7). Having a nonzero determinant, **R** is nonsingular and has an inverse. The same is also true for **M.** The inverse of **R** is

$$\mathbf{R}^{-1} = \mathbf{C}^T/|\mathbf{R}| = \begin{bmatrix} N & 0 \\ 0 & D \end{bmatrix} /|D.N| = \begin{bmatrix} 1/D & 0 \\ 0 & 1/N \end{bmatrix}$$

where $\mathbf{C}^T$ is the transpose of the cofactor matrix (matrix of cofactors) of **R**. As seen, the main diagonal elements of $\mathbf{R}^{-1}$ are the reciprocal of the corresponding diagonal elements of **R**. The same is also true for $\mathbf{M}^{-1}$. As **R** has an inverse, the following statements hold:

$$\mathbf{RR}^{-1} = \mathbf{I} \rightarrow \begin{bmatrix} D & 0 \\ 0 & N \end{bmatrix} \times \begin{bmatrix} 1/D & 0 \\ 0 & 1/N \end{bmatrix} = \begin{bmatrix} 1 & 0 \\ 0 & 1 \end{bmatrix}$$

$$\left(\mathbf{R}^T\right)^{-1} = \left(\mathbf{R}^{-1}\right)^T$$

That is, multiplying **R** with its inverse yields an identity matrix, and the inverse of its transpose is equal to the transpose of its inverse. Both of these are also true for **M** (see Sect. 7.5, Chap. 7).

The resemblance of **R** and **M** in such features can also be explained geometrically. **R** represents a vector space as **M** does. The difference is that the vector space represented by **R** is a rectangle (if $|D| \neq |N|$) or a square (if $|D| = |N|$), while the vector space represented by **M** is always a square—a DE square. The former is defined in the two-dimensional divisor-dividend vector space, and the latter in the two-dimensional debit-credit vector space. The determinant $|\mathbf{R}|$ in absolute terms is equal to the size of the ratio rectangle or square represented by **R**, and the determinant $|\mathbf{M}|$ is equal to the size of the DE square represented by **M**. As for the unit financial ratio in particular, **R** has a feature similar to the feature of **M**; it reflects the perfectly balanced relationship between the two financial items compared in the ratio, while **M** reflects the perfectly balanced relationship between the dual sides (debit and credit) of accounting events.

Summary and Concluding Remarks 10

The use of geometry to learn and study concepts and theories is common in many fields of study, including mathematics, physics, economics, and finance. Yet it is hardly the case in accounting. The book shows that the use of geometry is possible in accounting in order to present (record) accounting transactions (events) according to the double entry principle and explore the features and principles of double entry accounting. In doing so, the book uses geometric concepts and theories and explains things with various types of transaction examples. The geometric settings used to do so are two- and three-dimensional Cartesian spaces. The point, line, and space representing individual accounting events are referred to as the double entry (DE) point, DE line, DE square (or cube), and DE circle (or DE sphere) respectively of accounting events. For a given accounting event, these alternative geometric presentations are equivalent in that they all represent the same accounting event and reflect the same duality and balance features of the event or, more generally, the features and principles of double entry accounting.

Along with the geometric presentations, the book also shows how to present accounting events and explore the features of double entry accounting in multidimensional vector spaces, based on concepts and theories in linear algebra. In doing so, it uses the same accounting transaction examples mentioned above. The vector, and the matrix, representing individual accounting events is referred to as the double entry (DE) vector and DE matrix respectively of accounting events. For a given accounting event, both of these are equivalent and reflect the same duality and balance features of the accounting event, just as the DE point, DE line, and DE square (or cube) respectively of the accounting event do.

As an extension, the book also examines the aggregation of monetary measures in accounting. This subject is examined since the aggregation of those measures is part of the main task of accounting. As a further extension, the book also examines financial ratios in multidimensional geometric settings. This additional subject is examined since those ratios are generated with financial numbers, which come from accounting. These financial numbers are the aggregates of the monetary measures of individual accounts or groups of similar accounts, which are generated under the

© The Author(s), under exclusive license to Springer Nature Switzerland AG 2025

B. T. Ro, *The Geometry of Accounting*, Springer Texts in Business and Economics,
https://doi.org/10.1007/978-3-031-88892-2_10

double entry principle. The aggregation of monetary measures and the presentation of financial ratios are also examined in vector and matrix forms respectively in two-dimensional vector spaces.

The geometric approach adopted in this book is a notable departure from the traditional approach to accounting. Under the traditional approach, the presentation of accounting events and the features of double entry accounting are learned and studied in an accounting-only frame. In doing so, the traditional approach relies exclusively on numbers and technical terms unique to accounting such as debit, credit, double entry, journal, T-account, ledger, and so on. No geometric concepts or graphical tools are used. As such, the geometric approach in this book can be a new avenue for learning and studying accounting.

The book covers many aspects of accounting events and many features of double entry accounting, using concepts and theories in geometry and linear algebra. However, there are still certain issues that are yet to be examined geometrically. These are left out as subjects for future work. They are

(a) The quality of accounting events—actual versus estimated, discrete versus continuous, and verifiable versus unverifiable.
(b) The quality of accounting measures—actual versus estimated, discrete versus continuous, and verifiable versus unverifiable.
(c) The informativeness of accounting events.
(d) The quality of financial ratios—verifiable versus unverifiable.

The Quality of Accounting Events Accounting events (transactions), defined in Chap. 1, include not only arm's-length transactions between two parties but also any other event that affects the sources and uses of the capital resources of a business entity (firm). The sources and uses of the capital resources and their relationship are shown by a mathematical equation known as the fundamental accounting model: assets = liabilities + owner equity. Any event that affects the model or the relationship among its elements in one way or another is considered an accounting event. Thus, accounting events include not only arm's-length transactions between two parties but also other events that affect the elements of the model, such as the accrual of certain expenses or revenues, the depreciation of certain assets, and even a natural event such as earthquake, flood and hurricane. Given the broad definition of accounting events, not only are there different kinds of accounting events, but there can also be a wide range of differences in their quality.

Arm's-length transactions are actual, discrete, and verifiable so that they are of relatively high quality. However, other accounting events such as non-arm's-length transactions are not as high in quality since they are not actual, discrete, or easily verifiable. They are abstract or continuous and, hence, hard to verify or measure in monetary terms. Because of this, the monetary quantity of the debits and credits of such transactions is often estimated. Whether and how such differences in quality among different accounting events can be examined geometrically is an interesting issue for investigation.

The Quality of Accounting Measures All monetary measures in accounting originate from the debits and credits of accounting events, which come to exist as accounting events take place and are recorded. Hence, the quality of accounting measures depends largely on the quality of accounting events. This is true especially for the primitive measures of individual accounts as opposed to their derived measures (aggregates). The former are the debits and credits of individual accounts that are measured and recorded when accounting events involving the accounts are recorded. As the quality of accounting events varies, so are the quality of the monetary measures of those debits and credits likely to vary. In general, the primitive (debit and credit) measures of individual accounts relating to arm's-length transactions are of high quality since they are actual, discrete, and verifiable as the transactions are. Hence, they are likely to be more accurate, more objective, and more reliable than the primitive measures of individual accounts relating to accounting events that are not arm's-length transactions. For the latter transactions, the primitive measures of the debits and credits of individual accounts are abstract (estimated), continuous, or unverifiable in many cases. Given this, the quality of those primitive measures is likely to be not as high. How and whether such quality differences in monetary measures among different accounting events can be addressed geometrically is another interesting issue for investigation.

The Informativeness of Accounting Events Accounting events are informative in that they contain information about how they affect the debit and credit accounts involved and the sources and uses of the capital resources of a firm. Specifically, they tell what debit and credit accounts are affected and whether these accounts increase or decrease and by how much. They also give information about how the changes in those accounts affect the sources and uses of the capital resources of a firm and their compositions. The informativeness of accounting events as to such questions is likely to vary with their complexity (or simplicity) as well as quality.

Given the above, some questions arise: Are two accounting events having the same monetary size equally informative, all else equal? Are accounting events that are actual, discrete, and verifiable more informative than those that are not? Are accounting events having component transactions more informative than otherwise similar accounting events that do not have component transactions, when all else is equal? Or, more generally, is the *whole* more (or less) informative than its *parts* as a set when it comes to accounting events as in the case of the segment reporting, for example—the reporting of total sales of a firm by lines of business? Whether and how questions such as those above can be addressed geometrically can also be an interesting area for investigation.

The Quality of Financial Ratios The quality of financial ratios depends largely on the quality of the financial numbers used to generate them. These numbers are level 2 or level 3 aggregates of the monetary measures of individual accounts or groups of similar accounts. And those aggregates originate from the primitive monetary measures, namely debits and credits, of individual accounts, which are recorded when accounting events involving those accounts are recorded during an accounting period. Thus, the quality of those primitive measures depends on the quality of accounting events. As the quality of accounting events can vary widely, so can the quality of the primitive measures and hence their aggregates vary, which eventually affects the quality of financial ratios. Given this, not all financial ratios are likely to be the same in quality. Does the quality of the financial numbers used to generate financial ratios affect the quality of the ratios? If so, does it affect different financial ratios differently or in the same way? Does the quality of different financial ratios differ for a given firm and across different firms? How questions such as those above can be examined geometrically is another interesting area for investigation.

Index

© The Editor(s) (if applicable) and The Author(s), under exclusive license to Springer Nature Switzerland AG 2025

B. T. Ro, *The Geometry of Accounting*, Springer Texts in Business and Economics,
https://doi.org/10.1007/978-3-031-88892-2

The manufacturer's authorised representative in the EU is Springer
Nature Customer Service Centre GmbH, Europaplatz 3, 69115 Heidelberg,
Germany. If you have any concerns regarding our products, please
contact ProductSafety@springernature.com

Printed and bound by CPI Group (UK) Ltd, Croydon, CR0 4YY
30/09/2026
02225592-0008